FIRST BIENNIAL REPORT

OF THE

PROGRESS

OF THE

GEOLOGICAL SURVEY

OF MICHIGAN,

EMBRACING OBSERVATIONS ON THE

GEOLOGY, ZOÖLOGY AND BOTANY

OF THE

LOWER PENINSULA.

MADE TO THE GOVERNOR, DECEMBER 31, 1860.

By Authority.

LANSING:
Hosmer & Kerr, Printers to the State.
1861.

REPORT OF THE STATE GEOLOGIST.

To His Excellency MOSES WISNER,
Governor of the State of Michigan:

I have the honor to submit herewith, the Report required of me by the terms of the Legislative act, approved February 15, 1859, and entitled "An act to finish the Geological Survey of the State." This Report is intended to set forth the progress of the geological survey during the years 1859 and 1860.

Allow me, in communicating this Report, to acknowledge the many personal kindnesses received at your hands, and the appreciative interest which you have always manifested in the progress of the work. Whatever useful results may be here embodied, are due in no small degree to your connexion with the origin and energetic prosecution of the survey.

I have the honor to be,
Your most obedient servant,
A. WINCHELL,
State Geologist.

Ann Arbor, December 31, 1860.

INTRODUCTION.

SKETCH OF THE HISTORY OF GEOLOGY IN MICHIGAN.

Before entering upon the consideration of the subjects strictly belonging to this Report, a brief notice of what has heretofore been done in developing the Geology of Michigan, will undoubtedly be acceptable to the people of our State.

The explorations and discoveries of the Jesuit Missionaries, prosecuted for many years along the borders of the great Lakes, may be passed over as too remotely connected with the history of Geology in Michigan, to justify their introduction into the present report. The record of the wonderful labors and sufferings of these early christian missionaries, may be found embodied in the numerous volumes of a work entitled, "Relations de ce que s'est passe de plus remarquable aux Missions des peres de la compagnie de Jesus, en la Nouvelle France."* A condensed sketch derived from this source, is given in Foster and Whitney's "Report on the Geology and Topography of a portion of the Lake Superior Land District, in the State of Michigan, Part I."

The existence of copper in considerable quantity, upon the shores of Lake Superior, had all along attracted the attention of the Missionaries. The first mention made of the occurrence of this metal is found in the Relation for 1659–60. The first mining enterprise of which we have any account, was superin-

*A copy of this remarkable and rare old work is in the possession of Judge Campbell, Prof. of Law in the University. Other sources of information relative to this period are "Travels of the Jesuits into various parts of the world," &c. Vol. II., London, 1762. "Early Jesuit Missions in North America," by Rev. William Ingraham Kipp; New York, 1847. "Lettres edifiantes et curieuses," &c. Tome premier, pp. 637–818, Paris, 1846. For a knowledge of these works, I am indebted to Prof. White. Much further information may be found in the "Histoire de la Nouvelle France," and "Thevenot's Relations de divers Voyages Curieux," and "Recueil de Voyages," Paris, 1681.

tended by Alexander Henry, near the forks of the Ontonagon river, in 1771.

The explorations of Alexander McKenzie, commenced in 1789, extended over a portion of the shores of Lake Superior, and thence north-west, over the country whose waters flow into the Arctic ocean. In the account of his travels he speaks of the occurrence of "virgin copper" on the south shore of the lake.

In the year 1800, during the presidency of the elder Adams, Congress passed a resolution,* providing for the employment of an agent for the purpose of collecting information relative to the "Copper Mines" on the south shore of Lake Superior; but it does not appear that this resolution was ever put in execution.

In 1819, General Cass, under the authority of the Secretary of War, directed an exploring expedition which passed along the Southern shore of Lake Superior,† and crossed over to the Mississippi. This expedition had among its principal objects, that of investigating the north-western copper mines; and was accompanied by Mr. H. R. Schoolcraft in the capacity of mineralogist and geologist. His observations are recorded in his "Narrative Journal of Travels from Detroit, north-west," &c., published in 1821.

In the spring of 1823, Major Long, acting under the orders of the War Department, and accompanied by several scientific gentlemen, started on an expedition, the object of which was to explore the river St. Peters and the country situated on the northern boundary of the United States, between the Red River of Hudson's Bay, and Lake Superior. In returning, they coasted along the north shore of this Lake.

In 1831 an expedition was sent out by the United States government, under the command of Mr. Schoolcraft, for the purpose of ascertaining the sources of the Mississippi river. Dr. Douglas Houghton was attached to this party, and he subsequently

*Laws of the United States, Vol. III., p. 403.
† Journal of the Expedition of General Cass.

speaks of the aid afforded by the observations made at this time in tracing the fragments of copper to their place in the rock.

Nothing further was attempted at elucidating the mineral resources of any portion of the territory, until the admission of Michigan into the Union in 1836, when the government at once proceeded to the organization of a general, systematic survey.

The original act for the organization of the geological survey of the State was approved by Gov. Mason, February 23d, 1837.* It provided for a geological, zoological, botanical and topographical survey. Under this act the following corps of officers was appointed.†

Douglas Houghton, Geologist.

Abram Sager, Principal Assistant, in charge of Botanical and Zoological Departments.

S. W. Higgins, Topographer and Draughtsman.

Columbus C. Douglas, Sub-Assistant.

Bela Hubbard, Sub-Assistant.

William P. Smith, Sub-Assistant in charge of Mechanical Zoology.

Messrs. Douglas and Hubbard, during the following years, were First Assistants.

On the 26th of January, 1838, Dr. Houghton presented his *First Annual Report*, a document of 37 pages, in which, after alluding to the topography of the State, he notices the several geological features of the Lower Peninsula under the following heads: "*Upper Sandrock of the Peninsula,*" "*Gray Limestone,*" "*Lower Sandstone or Graywack Group,*" "*Gypsum,*" "*Brine Springs,*" "*Clay,*" "*Sand,*" "*Marl,*" "*Bog Iron Ore,*" "*Mineral Springs.*" Several pages are devoted to the Brine Springs, and numerous interesting analyses of the saline waters of Michigan are for the first time published.

On the 22d of March the Governor approved a new act, re-organizing the survey in more comprehensive terms, and with more detailed provisions.‡

* Senate Journal 1837, p. 189. For the Act, see "Laws of Michigan," 1837, p. 14.
† Report, 1838.
‡ "Laws" 1837-8, p. 119.

About the same date, acts were passed for the incorporation of the "Clinton Salt Works," and for the improvement of the State Salt Springs."* January 1, 1839, Dr. Houghton presented a special "Report in relation to Salt Springs,"† and on the 28th of the same month, a Report on Iron Ore in Branch County.‡ The same day the Legislature passed an "Act relative to Salt Springs."

On the 4th of February, 1839, Dr. Houghton presented his "*Second Annual Report.*"§ This document, of 153pp., was made up as follows :

1. Geology, by Dr. Houghton, 39 pp. "Northern Part of the Peninsula." "*Topography and General Character,*" "*Rocks,*" "*Tertiary Clays,*" "*Shell Marl,*" "*Gypsum,*" "*Change of Elevation in the Waters of the Great Lakes.* Southern Part of the Peninsula. "*Coal,*" "*Salt Springs and State Salt Lands.*"

2. Zoology, by Dr. Abram Sager, 15 pp. A systematic catalogue of the animals of the State, as far as observed.

3. Botany, by Dr. John Wright, 29 pp. A catalogue of the plants of the State as far as observed.

4. Topography, by S. W. Higgins, 21 pp.

5. Geology of Eaton, Ingham and Jackson counties, by C. C. Douglas, Assistant Geologist, 13 pp.

6. Geology of Wayne and Monroe counties, by Bela Hubbard, Assistant Geologist, 36 pp.

The Zoological and Botanical Departments were suspended early in the year by the resignation of the officers in charge.

On the 6th of January, 1840, the State Geologist made a report in relation to the Salt Springs‖, and on the 3d of February, presented his *Third Annual Report*¶ of 111 pages, covering the following documents :

1. Geology, by Dr. Houghton, 33 pp. A description of the Topography and Geology of that portion of the Upper Penin-

* Laws 1837-8, p. 165.
† House Doc., p. 39.
‡ Ib., p. 342.
§ House Doc., 1839, p. 380.
‖ House Doc., 1840, Vol. I, p. 13.
¶ Ib. Vol. II, p. 202.

sula bordering on Lakes Michigan and Huron, followed by a notice of the "Clay, Iron Stone and Bog Ores" of the Southern Peninsula. The rocks of the Upper Peninsula are here arranged under the two heads PRIMARY and SEDIMENTARY. The latter are subdivided into *Lower Limerock and Shales*, and *Upper Limerocks.*

2. TOPOGRAPHY, by S. W. Higgins, 18 pp.

3. GEOLOGY, by C. C. Douglass, 23 pp., containing "General Remarks on the counties of Jackson, Calhoun, Kalamazoo, Eaton, Ionia and Kent," with considerable detail on the rocks of the coal measures, which are divided into "Upper" and "Lower Coal Groups."

4. GEOLOGY, by B. Hubbard, 35 pp. containing reports on Lenawee, Hillsdale, Branch, St. Joseph, Cass, Berrien, Washtenaw, Oakland and Livingston Counties, and embracing a systematic description of the various formations and economical products of these Counties ; a notice of the "Ancient Lake Ridge," and numerous practical suggestions on the use of Peat and Marl.

A Committee of the House of Representatives reported on the reports of the State Geologist, at this session of the Legislature,* and the Zoological and Botanical portions of the act establishing the Survey were repealed. On the 28th of March, an act was passed relative to the maps of the State and Counties.

The *Fourth Annual Report* of the State Geologist was presented February 1, 1841. This Report embraced the following documents :

1. GEOLOGY, by Dr. Houghton, 89 pp. This was devoted to a description of the Topography, Geology and Minerology of the country bordering on Lake Superior. The classification of the rocks will be embraced in the table which follows. The report embraces a masterly discussion of the Mineral Veins of the "Trap, Conglomerate, &c.," and concludes with notices of the "Furs, Fish and Harbors of Lake Superior."

2. LATITUDES AND MAGNETIC VARIATIONS, by Frederick Hubbard, Special Assistant, 6 pp.

*House Doc. 1840, Vol. II. p. 455.

3. Geology, by C. C. Douglass, 15 pp., devoted mainly to the general geology of the northern portion of the Lower Peninsula, bordering on Lakes Huron and Michigan. The geological series, as here made out, will also be embraced in the table.

4. Geology, by B. Hubbard, 33 pp., devoted to a general *resume* of the geology of the organized counties, with tables of the formations.

5. Topography, by S. W. Higgins, 26 pp., containing valuable tables of magnetic variations, and of the rise and fall of water in the lakes.

On the 4th February, Dr. Houghton presented a Report of the progress of the County and State maps.*

Through the pressure of the financial crisis under which the State and country were still suffering, the Legislature was induced to curtail the appropriations for the continuance of the survey. The *Fifth Annual Report* therefore, dated January 25th, 1842,† is a brief paper of six pages, containing some notices of the geology of the western portion of the Mineral District of Lake Superior, surveyed by Dr. Houghton in connexion with his duties as Boundary Commissioner. Dr. Houghton, not content that a work to which he had devoted so much labor, and for which he had undergone so many privations, should be interrupted, and perhaps frustrated, by the supposed inability of the State to carry it on, devised, in 1844, in connexion with William A. Burt, Esq., the plan of connecting the linear surveys of the public lands of the United States, with a a geological and mineralogical survey of the country. This plan was fully set forth in a paper prepared and read by him before the "Association of American Geologists," at Washington, in that year. The immense advantages likely to result from such a survey, if successfully carried into execution, were at once comprehended. The Commissioner of the General Land Office, having obtained a promise from Dr. Houghton to undertake the work, recommended to Congress an

*House Doc., 1841, p. 94.
† Joint Doc., 1842, p. 436.

appropriation for that purpose. This was made, and the survey commenced by Dr. Houghton.* According to the plan agreed upon between Dr. Houghton and Mr. Burt, the township lines of the Upper Peninsula were to be run by Mr. Burt, or under his supervision, while the subdivisions were to be made by other deputy surveyors—Dr. Houghton having the especial control of the whole. All rocks crossed by lines were to be examined, specimens taken, and the exact locality noted, while at the same time as much information as could be obtained, was to be collected in relation to the geological and topographical features of the country. The surveyors were to be accompanied along the lines by a special barometrical observer. This system had been fairly organized, and the field work of one season nearly completed, when his melancholy death by drowning, occurred during a storm on Lake Superior, near Eagle river, on the night of Oct. 13th, 1845.† This unfortunate termination of the survey was communicated to the Legislature by S. W. Higgins, on the 7th of January, 1846.‡

According to the plan entered upon, a full and minute report was to have been prepared and returned by Dr. Houghton, to the office of the Surveyor General. On the decease of the head of the survey, his administrators employed Messrs. William A. Burt and Bela Hubbard, to compile reports on the geological results of the work for 1845, from the field notes of that year. Mr. Burt's Report was prepared from his own notes, and Mr. Hubbard's from those of Dr. Houghton. These two Reports§ unfold in an admirable manner the geological structure of the trap and metaphorphic regions of Lake Superior, and anticipate results which were subsequently worked out by the United States Geologists. The notes and maps of three townships were in Dr. Houghton's possession at the time of his death, and were never recovered.

Thus ended the first geological survey of our State—a work

* See "The Mineral Region of Lake Superior," by Jacob Houghton, Jr.
† Ib. Also Foster and Whitney, Rep. Vol. I, p. 14.
‡ Joint Doc. 1846, No. 12.
§ For my knowledge of these Reports I am entirely indebted to the work of Jacob Houghton, Jr., before referred to.

inaugurated within a little more than a year after her admission into the Union, and prosecuted, consequently, in the midst of the greatest embarrassments. But though the work was unavoidably arduous for the geologist, and expensive for the State, it served to acquaint the world, at an early day, with many of the sources of our mineral wealth, and to awaken and maintain a lively desire for more full and definite information relative to the Coal, Salt, Gypsum, Copper and Iron, of which the published Reports of Progress had afforded hasty glimpses. Dr. Houghton's Report, published in 1841, furnished the world with the first definite information relative to the occurrence of native copper in place, on Lake Superior;* and the promise of wealth now so rapidly growing up in that region, has been to a great extent created by the attention drawn in that direction by this Report of my lamented predecessor.

The subjoined table, setting forth the order of arrangement of the rocks of the State, as compiled from Dr. Houghton's Annual Reports, and those of his assistants, will perhaps sufficiently extend, for the present occasion, this historical reference to the former State Geological Survey.

Succession of Strata in Michigan, as published in 1838–41, *Arranged in Descending order.*

XXXI. Recent Alluvions, (Hubbard, Rep't 1841, p. 122.)

XXX. Ancient Alluvions, (Ib. 120.)

XXIX. Erratic Block Group or Diluviums, (Ib. 115.)

XXVIII. Tertiary Clays. (Houghton, 1839, p. 17; 1841, p. 43; Hubbard, 1841, p. 123.)

XXVII. Brown or Gray Sandstone. (Douglass, 1840, p. 69; Hubbard, 1841, p. 130.)

XXVI. Argillaceous Iron Ore in thin included beds, (Ib.)

XXV. Coal Strata, alternating with friable, slaty sandstone, and thick beds of black shale and slate, (Ib.)

XXIV. Red or variegated sandstone. (Douglass, 1840, p. 70; Hubbard, 1841, p. 129.)

*Whitney's Metalic Wealth of the United States, p. 243.

XXIII. Gray and yellow sandstone. (Hubbard, 1841, p. 128.)

XXII. Shales and coal of the "lower coal measures." (Douglass, 1840, p. 65; Hubbard, 1841, p. 126.)

XXI. Blue, compact, slaty sandstone. (Hubbard, 1841, p. 136.)

XX. Gray limestone, or Upper Limerock—14 ft. (Hubbard, 1841, pp. 125–130.) Douglass places this between the "Upper" and "Lower Coal," and says the Gypsum is above, or embraced in it. (1840, pp. 62–67.) The Gypsum is also placed above by Houghton. (1839, p. 11.)

XIX. Fossiliferous ferruginous sandstones. (Hubbard, 1840, pp. 81–88.) Thinned out at Grand Rapids. (Hubbard, 1841, p. 138.) Subdivided as follows:

G. Coarse, quartzose, yellowish gray sandrock, 30 ft.

F. Ash colored, or brown sandrock, with marine fossils, 15 ft.

E. Dingy and green, finegrained strata, with occasional fossils and ferruginous spots.

D. Hard gray stratum of sandrock, 1 ft.

C. Dingy-green, finegrained sandstone, interstratified with slaty sandstone, and apparently with blue clay shale, 15 to 20 ft.

B. Yellow, fossiliferous sandrock. Abounds in marine fossils. 20 ft.

A. Finegrained sandrock.

XVIII. Kidney Iron Formation, 45 ft. (Hubbard, 1840, p. 86; 1841, p. 13; Houghton, 1840, p. 25.) Considered the bottom of the Carboniferous System.

XVII. Sandstone of Pt. aux Barques. Passes south-west and underlies the sandstone of Hillsdale county, though not exposed there. (Hubbard, 1841, p. 132.) Divided as follows:

B. Coarse sandstone or partial conglomerate. (Hubbard, 1841, p. 136.)

A. Yellow and greenish sandstones. (Ib.) The sandstones XVII are supposed to be equivalent to the Ohio "Conglomerate" and "Waverly Sandstone." (Hubbard, 1841, p. 132.)

XVI. Clay Slates and Flags, of Lake Huron, 180 ft. (Hubbard, 1841, p. 136.) Divided into

B. Argillaceous sandstone, alternating with sandstone and clay slates.

A. Blue clay slates and flags, with alternating gypsum beds and gypseous marls. These two (A and B) constitute the "Upper Salt Rock" (Hubbard, 1841, p. 133). The gypsum of Grand Rapids is placed here by Hubbard, (1841, p. 133).

XV. Pt. au Gres and Manistee* limestone (Douglass, 1841, pp. 102, 103).

XIV. Soft, coarse-grained sandstones, 230 ft. (Hubbard, 1841, p. 133.) Pierced at Grand Rapids, in the salt well of Lucius Lyon. The "lower salt rock" of Ohio, Va. and Mich. (Ib. 133.)

XIII. Black bituminous, aluminous slate, with pyrites (Hubbard, 1841, p. 134).

B. Light blue, argillaceous (Douglass, 1841, p. 102).

A. Black, containing pyrites. (Ib.)

XII. Limestone of Lake Erie, (Hubbard, 1839, pp. 88, 105; 1840, p. 83; 1841, p. 134). Subdivided as follows:

D. Corniferous limestone, (Douglass, 1841, p. 102).

C. Thunder Bay and Little Traverse Bay limestones, (Douglass, 1841, pp. 112, 103).

(*f*) Blue silicious limestone, (Douglass, 1841, p. 109.)

(*e*) A confused mass of broken fossils, imbedded in clay. (Ib.)

(*d*) Vesiculated chert, colored with iron. (Ib.)

(*c*) Flaggy limestone in very thin layers. (Ib.)

*The limestone referred to by Douglass seems to be rather on the Muskegon than on the Manistee river, which is many miles further north.

(*b*) Blue clay with iron pyrites. (Douglas, 1841, pp. 109.)

(*a*) Light blue limestone.

B. Black bituminous limestone. (Douglas, 1841, pp. 102, 103.)

A. Blue limestone. (Ib.)

XI. Mackinac limestone, (Douglas, 1841, p. 102, 103,)—"Manitoulin Portion of Upper Limerock." (Houghton, 1840, pp. 19, 21.)

X. Polypiferous Portion of Upper Limerock. (Houghton, 1840, pp. 19, 21.

IX. Pentamerus Portion of Upper Limerock. (Ib.)

VIII. Lower Limerock and Shales. (Houghton, 1840, p. 16.)

VII. Sandy Limerock. (Houghton, 1841, p. 20.)

VI. Upper grey Sandstone. (Houghton, 1841, p. 19.) Not conformable with next stratum.

V. Lower, or Red Sandstone and Shales. (Houghton, 1841, p. 119.)

IV. Mixed Conglomerate and Sandstone. (Ib.)A

III. Conglomerate. (Ib. 17.)B

II. Metamorphic Rocks. (Ib. 16.)C

I. Primary Rocks. (Ib. 15)D

Little more than a year after the suspension of the survey under Dr. Houghton, Congress passed an act, approved March 1st, 1847, embracing provisions for the geological exploration of the Lake Superior Land District, organized by the same act. Under this act, Dr. C. T. Jackson was appointed by the Secretary of the Treasury, to execute the required survey.

After having spent two seasons in the prosecution of this work, he presented a report of 801 pages,* and resigned his commission. In the meantime, the survey was continued, and subsequently completed by Messrs. Foster and Whitney, United States Geologists. Their Report, of 224 pages, on the "Copper

---

A, B, C. D. The Traps intersect this series variously.

*Ann. Mess. and Doc. 1849–50, Part III. Also, Senate Doc. 1st Sess. 31st Cong. Vol. 3, 1849–50.

Lands," was submitted as Part I., on the 15th of April, 1850.* Part II., on the "Iron Region" and General Geology, was submitted November 12th, 1851, and forms a volume of 406 pages and XXXV. Plates.† Messrs. Foster and Whitney were aided in the field work of the survey by Messrs. S. W. Hill and Edward Desor as first assistants; by Mr. William Schlater as Draughtsman, and Mr. W. D. Whitney as Botanist. The fossiliferous region was also passed over by Prof. James Hall, the palæontologist of New York, whose observations and general conclusions are embodied in the Report, together with papers on the Geology of Wisconsin, by Dr. I. A. Lapham, and Col. Chas. Whittlesey. The latter also communicated important chapters on the "observed fluctuations of the surfaces of the Lakes," and "magnetic variations," with a "comparison of terrestrial and astronomical measurements."

The examinations reported upon in Part II., extended around the entire Lake shores of the Upper Peninsula, as far as the head of Green Bay, and embraced the islands at the head of Lake Huron, from Mackinac to Drummond's Island. The groups of of rocks observed were found to conform to the geology of New York and other States, and a parallelism was established, from the Potsdam Sandstone to the Upper Helderberg Group.

No further public geological explorations were made within the limits of our State, until the commencement of the present survey. The mining companies of Lake Superior, however, maintained a series of local explorations, which have contributed a vast amount of detailed information, destined to be of the greatest service in the compilation of a general report.

During the legislative session of 1858, numerous petitions were presented for the completion of the geological survey of the State. The number was greatly increased at the session of 1859, and, although the condition of the State Treasury was reported to be such as hardly to justify embarkation upon any extraordinary expenditures, it was finally deemed advisable to make

*Executive Doc. No. 69, 1st Sess. 31st Cong., Vol. 9, 1849–50.
†Executive Doc. No. 4, Special Sess. 32d Cong., Vol. 3, 1851.

a commencement of the work of completing the geological survey, and preparing for publication in a convenient and practical form, a Report upon the Geology of the State, drawn from original observations, and all other accessible sources. The terms of the act approved February 15, 1859, were copied almost literally from the original act of 1837 ; and the survey thus instituted, possessed, of course, all the scope of the original undertaking.

The following Report embraces only the results of the actual observations made during the past two seasons. It must be borne in mind, however, that the provision made for the prosecution of this survey, has not as yet been such as to permit its complete organization ; and the heads of the principal departments have only been employed during such time as could be spared from other and regular professional engagements.

PART I.

GEOLOGY.

CHAPTER I.

ORGANIZATION OF THE SURVEY, AND PLAN OF OPERATIONS.

On the receipt of my commission, dated March 9th, 1859, designating me to take the charge of the work provided for by the "Act to Finish the Geological Survey of the State," approved February 15, 1859, I met you, by request, for the purpose of consultation upon a plan of operations adapted to the circumstances then existing. Besides the act just referred to, a joint resolution had been previously passed, making an appropriation for the "publication of Dr. Houghton's Notes." An examination of such sources of information as were accessible, had shown, however, at the time of our interview, that there were no "notes" made by Dr. Houghton or his assistants which had not already been reported from, either by the observers themselves, or in the manner provided for by Dr. Houghton's administrators, except the field notes upon four townships in the Upper Peninsula, which were lost at the time of the melancholy occurrence which terminated the original survey. It resulted, that the only further use which could be made of the "notes" referred to in the resolution, would be to work them up into a detailed report upon the geology of the State, as understood twenty years ago, before the geology of New York, Canada, and the Northwest, had contributed such important aids to the proper understanding of the geology of our own State. While such a publication as this would be but a just tribute to the men who had labored and suffered for years in this great work, it was not deemed compatible with the interest of the State, nor conducive to the advancement of science, to prosecute the publication of Dr. Houghton's "notes" in all the details of a final report, and immediately follow it with another report, already provided for, which should complete the elucidation of our geology, and adapt it in all respects to our present wants and the present state of the science. This view seemed the more consistent, since any adequate report upon our geology could not fail to do justice to the names of those who were the pioneers in Michigan geology.

After the interval which had elapsed since the date of the explorations made in the Lower Peninsula of the State, it was obvious that a great multitude of facts must have come to light, calculated to have a bearing upon any final conclusions as to the geological succession of our strata. New natural exposures of the underlying rocks, had been discovered, new quarries had been opened, the working of coal and gypsum had actually commenced on a successful scale; and especially were new opportunities presented for the collection of fossils—the language in which geological records are written. It seemed necessary, therefore, to undertake the same thing which had been undertaken by Massachusetts, by South Carolinia, Tennessee, and other States. The ground was to be gone over again, for the purpose of posting up our collection of facts. The Lower Peninsula, as being least understood, was to receive the first reconnoissance. Detailed examinations were to be made, only with reference to settling the geographical distribution of the coal, and resolving other questions of immediate economical importance. The report which follows, will show I trust, a satisfactory degree of success in making these determinations. The geological observers, in the progress of their work, were to embrace favorable opportunities for the collection of zoological and botanical specimens; and the zoological observers were to make note of all geological data which came in their way.

In pursuance of the plan agreed upon, I made an excursion on the first of April to the Maumee river, where, by the enlightened liberality of Mr. George Clark, the proprietor of several fishing stations on the rivers and lakes, I was enabled to secure two or three barrels of specimens of the various species of fish, and other aquatic animals common to south eastern Michigan and northern Ohio. At the same time, Dr. Manly Miles, who had been designated to take special charge of the department of Zoology, descended the Saginaw river to its mouth, in company with a young man who was subsequently employed during the season as taxidermist and general assistant. About the middle of May, the necessary preparations having been completed, I entered upon the field work of the season, by commencing a geological survey of the county of Monroe. In this part of the work I was accompanied by Messrs. A. D. White and Lewis Spalding, two students of the University, who volunteered their assistance for the mere payment of their traveling expenses. Mr. White continued in the service of the survey during the season, and was again employed the present season. I am happy here to testify to the faithful, able and obliging manner in which he has co-operated in the execution of all my plans.

After the completion of our observations in Monroe county, our reconnoissance was extended through Jackson, Hillsdale, Lenawee, Branch and Calhoun counties. Having familiarized myself with the character of the Coal Formation in the vicinity of Jackson, and traced its limits to the east and west of the city, I had no hesitation in pronouncing upon the non-existance of coal at Jonesville, or in Hillsdale county. I subsequently had the opportunity to discourage the explorations for coal in the vicinity of Albion, misguided to the same extent as those of Hillsdale county. Similar duties, always unpleasant, and often met by ingratitude and incredulity, have had to be performed in scores of other places. The observations made at Jackson, Woodville, Barry, Albion, Marshall, Battle Creek, Union City, Jonesville and Hillsdale, have proved exceedingly instructive, as will be shown in the ultimate publication of the details of the survey.

The south-western part of the State promising to be less productive of useful observations, Mr. White was instructed to traverse the counties of St. Joseph, Cass, Berrien, Van Buren and Kalamazoo, along designated lines, while I entered upon the examination of the regions bordering upon the lines of public conveyance. Our party of two, was thus converted into two parties, each attending, more or less, to all the departments of the survey. In the beginning of autumn, we met by appointment, at Grand Haven, and proceeded over the country to Grand Rapids. Here I made an examination of the geological relations of the gypsum and salt, and announced, as is believed, for the first time, the true geological position of those important products. Here Mr. White was detained several weeks by an intermittent, contracted from exposure at Grand Haven. In the mean time, however, he succeeded in making several excursions into the northern part of Kent county. Towards the last of October, I returned to Grand Rapids, and after completing my geological observations, communicated, by request, to James Scribner, Esq., in writing, my conclusions as to the geology of the Grand River Valley, and the depth at which the brine horizon would be found to lie. I stated that the source of the brine was from the shales of the gypseous group, near its base; and that I had no evidence of the existence of stronger brine at any greater depth in the formations which outcrop in the southern part of the State. I said that though the underlying formations are all somewhat saliferous, they are not strongly so, but that there are fissures and powerful currents of water at certain points, which would render extremely unpromising the search for salt below the gypsum formation. I recall these declarations at this time, for the purpose of vindicating the reliability of geological inductions, however unfavorable to individual or local interests and prejudices.

From Grand Rapids I proceeded to a cursory examination of the coal of Shiawassee county, and the brine of Saginaw county, while Mr. White proceeded through Barry, Eaton and Jackson counties, to Ann Arbor. I found the salt boring at East Saginaw progressing successfully under the enlightened management of Dr. Lathrop, one of the best geologists in our State, who had stimulated this enterprise as an inference from purely geological data. My observations upon the outcrops of the rocks which this boring was penetrating, enabled me to predict with considerable confidence the depth at which the salt bearing rocks of Grand Rapids would be reached. The subsequent result very happily justified my judgment, and we are enabled to day to look upon one of the leading enterprises of the State as the direct offspring of theoretical geology.

Dr. Miles, after spending some time on the Flint and Saginaw rivers, visited the numerous lakes and streams of Oakland, Livingston, Genesee, Lapeer and Washtenaw counties, and toward the close of the season paid a visit to the western part of the State. He was accompanied during most of the season by Mr. Dodge, of Flint, and for a few weeks by Dr Jewell, of Ann Arbor. While in the vicinity of Flushing, in Genesee county, he collected valuable observations and specimens from the outcrop of the coal series; and these have been communicated, with proper diagrams, to this department.

In December, I entered upon a comparative examination of Michigan and Ohio gypsum. A chemical analysis of each was made, at my request, by Prof. L. R. Fisk, of the Agricultural College; and at the instance of C. A. Trowbridge, Esq., of Detroit, I drew up a paper on the subject, which was placed in his hands.

In the month of February, 1860, I paid a visit, by request, to the salt works at Grand Rapids. I collected information, and explained the indications, in the light of the geological observations which I had made in the southern part of the State ; and while there, delivered a public lecture upon the subject of *Salt and its Geological Relations.*

About the first of March, I drew up a paper embracing a brief exhibit of the geology of the southern peninsula, and a condensed statement of the borings at various localities for salt. This was transmitted to Dr. Potter, the Superintendent of the salt operations at East Saginaw, for the purpose of informing the company which he represented, as to the geological position which their salt boring had reached, and what might be expected as the consequence of continuing to greater depths. For this communication, the company returned me a very polite resolution of thanks, with a request for permission to make the communication public.

On the 9th of April, I transmitted to you, an Informal Report on the progress and results of the survey, which, while not called for by the terms of the Act, was intended as an acknowledgement of the enlightened interest which you had all along manifested in the progress of this work.

Before the close of March, I had commenced preparations for the field work of 1860. On consultation with Dr. Miles, it had been agreed to unite the geological and zoological parties, and thus incur the expense of but a single outfit. The principal part of the season's business was to be upon and near the shores of the great Lakes. In these situations, where natural sections are always presented down to the surface of the water, rock exposures are much more frequent than in the interior. By determining the points on each side of the Peninsula, where the various formations intercept the lake shores, there is little difficulty in tracing approximately the lines of outcrop across the State.

It was intended to prosecute, before the season was sufficiently advanced for safe navigation in small boats upon the lakes, an examination of the vallies of the Cass and Tittibawassee rivers. Reports which had been rife during the previous season, of discoveries of coal, lead, iron, and "volcanic" rocks and "craters," in the vicinity of the Cass river, excited the hope that some unexpected developments might accrue from a scientific examination of that region; while on the other hand it was hoped that the ascent of the Tittibawassee would result in some revelations as to the nature and limits of the coal and salt formations. On the 18th of May, Dr. Miles and Mr. White set out upon the exploration of the Cass, but the anticipated survey of the Tittibawassee, by myself, was prevented by extreme family affliction, and death, occurring on the very day that I had designated for my departure.

Finding that a suitable boat could not be procured in the

lower part of the State without great sacrifices, I visited the Sault, in the last of May, and purchased, at a great saving, a Mackinaw boat, which proved to answer our purpose perfectly. Early in June, the surveying corps made a rendezvous at East Saginaw. Besides Dr. Miles and myself, the party consisted of Mr. White, Mr. N. H. Winchell, who had been engaged as botanical assistant, and two *voyageurs*, who had been secured at the Sault. After carefully exploring the whole coast, from the mouth of the Saginaw river to the vicinity of White Rock, the party returned and entered upon the examination of the west coast of the Bay and Lake, which was continued to Mackinac. After an examination of this and the neighboring islands we coasted along eastward to Drummond's Island, which, at this time and subsequently, was completely circumnavigated. We proceeded thence to the Bruce and Wellington mines, and thence to the Sault. Here my plan of operations called me to another part of the State; and as Dr. Miles, who had left the party at Thunder Bay on the 4th of July, did not rejoin it at the Sault, Mr. White led the explorations for the remainder of the season. His instructions took him back to the islands at the head of Lake Huron, and thence to Mackinac. At this place Dr. Miles rejoined the party. Thence they coasted along as far as Northport, on the south side of Grand Traverse Bay, following all the indentations of the coast, and entering all the small lakes accessible by navigable streams. From this point the party returned home.

In August I made an excursion to Cleveland and the Cuyahoga Falls, for the purpose of procuring data with which to compare my observations upon the coal measures and other perplexing strata in our own State.

In September, I made, by request, a special examination of the coal measures located in the vicinity of the Detroit and Milwaukie railway, in Shiawassee county, and transmitted the results of my observations to W. K. Muir, Esq., Superintendent.

I subsequently revisited Grand Rapids, and made examinations of some portions of Ionia, Clinton and Ingham counties.

Deeming that a popular exhibition at the State Fair of the economical results of the survey, thus far attained, would conduce to the diffusion of information relative to our resources, and the awakening of increased interest in the survey, I made for this purpose a selection from such specimens as had been at that time unpacked, and drew up, on a large scale, an outline map of the geology of the State to accompany the specimens in the exhibition. This undertaking appeared to be highly appreciated, the vicinity of the collection being continually crowded with interested observers and inquirers. The appreciative notices of the press were also of a very gratifying character. I have reason to believe that the exhibition, though very hastily got together, and very incomplete, was productive of considerable good.

In September, 1859, I issued a circular addressed to County Surveyors, and others throughout the State, the object of which was to procure reports from competent persons, on the topography of the various counties; the localities of rocks and minerals; the nature of the soil; the distribution of timber, &c. Several responses were promptly made to this circular, and I feel confirmed in the opinion that the county surveyors or former surveyors of the State, or in case they will not act, the private surveyors and engineers of the various counties, have it in their power to contribute to the prosecution of the geological survey, some of the most valuable information. Localities of rock exposure must almost always come to the knowledge of the linear surveyor, and, by communicating this knowledge to the geologist, great expense and delay may be saved, in traversing territory barren of geological indications. I would take this opportunity to urge upon surveyors and others, the importance of the service they are thus able to render to the geologist, with very little extra trouble to themselves.

The only communications actually received to this time in response to the above circular, are the following:

1. Kent county, and the region west and north-west. By John Ball, of Grand Rapids.

2. The valley of the Au Sauble river. By S. Pettibone, of Ann Arbor.

3. Brownstown, Wayne county. By B. F. Woodruff, of Brownstown.

Several other communications are promised, and supposed to be in progress.

The unpacking and labeling of the immense numbers of geological specimens required to illustrate the geology of all parts of the State, and complete the suites of duplicates called for by the Act establishing the survey, forms no inconsiderable share of the mechanical labor imposed upon the geologist The locality of each individual specimen must be preserved from the time it is broken from the rock, through all the vicissitudes of bagging, transportation by hand, boxing, transportation by public conveyance, and unpacking; and not only this, for where a cliff presents two or more strata successively superimposed, it is essential for the geologist to know what fossils or other specimens are afforded by each stratum. Allusion is here made to the subject, for the purpose of explaining thus early, the system of permanent labels which has been adopted. Every locality visited by the geological surveyors is designated by a separate number. These locality-numbers form a series reaching from the beginning to the end of the survey. In a book of localities provided for the purpose, the precise locality corresponding to each number is stated in full, to which are added the name of the owner of the land, (when known,) the formation exposed, the fossils found, and remarks. On every specimen collected is stuck a small oval piece of yellow paper on which is written the number designating the locality, which, in this way, is sure to be made an inseparable part of the specimen. The successive strata at any locality are designated by the letters of the alphabet, in all cases beginning at the lowest stratum.

The specimens collected during the past season have filled over a hundred boxes, and when it is known that each box contains from fifty to one hundred specimens, some idea may be formed of the amount of manipulation required for the permanent and effectual labeling of the specimens. The subsequent study of the specimens is still an additional labor.

Besides the keeping of the book of localities, every observer keeps a minute account of all his observations, written in a field book on the occasion, while the objects are before him and all their relations are fresh in his mind. Such inferences as the state of facts is calculated to suggest, are put down at the same time. Thus, though subsequent observations may materially modify or reverse these conclusions, they at all times possess the value of being the impression made upon the judgment, with all the observed facts vividly before the mind. All these notes are, at the end of the season, transcribed in order, in a Note Book kept for the purpose.

The third book kept is intended to show the geology of each township of the State. Under the several counties are arranged the townships in alphabetical order; and opposite the name of each, are references to every locality visited in it. Turning to these localities in the Note Book, all that has been learned of the township is at once before the eye.

Still another book is provided for memoranda, historical data, office work, &c. Thus, by this extensive and minute system of records and references, everything which has been done or learned is at all times immediately accessible; and no casualty to the geological corps, could result in losses as serious as when a large part of the data are left till the close of the survey, in the custody of individual memories.

The limited provision made for the prosecution of the survey, has rendered it impossible to engage the services of a chemist and mineralogist. The work of a geological survey—not including the zoology, botany, meteorology, and other researches generally attached to it—embraces field observations, collection of specimens, palæontology, mineralogy and chemistry; and it is seldom that a single person is competent to do requisite justice to all these departments. It is always desirable, therefore, to attach to the survey some suitable person to devote himself to the chemical examination of minerals, rocks, ores, soils, mineral waters, &c. This part of our survey has thus far been neglected. A few analyses have been made at my request, by Prof. L. R. Fisk, of the Agricultural College, by which that institution became connected with the survey, before the appointment of Dr. Miles to the chair of Zoology. By my arrangement with Prof. Fisk, he has not as yet received any compensation for his services, having agreed to await the action of the Legislature, in reference to further provision for the survey.

Immediately on the organization of the survey, I took steps to ascertain whether some portion of the scientific investigations might not be completed by experts of this and other States, who would, in many cases, expect no further compensation for their services than the opportunity of looking over our specimens, with permission to retain for their own cabinets, duplicates of such species as might prove to be novel or peculiar. I have accordingly had the satisfaction of being assured that different specialists stand ready to take up the different orders of our insects, and to furnish catalogues as soon as the specimens are placed in their hands. The same is true of some branches of the palæontology. Dr. H. A. Prout, of St. Louis, is already at work upon our Bryozoa, an important class of fossil mollusca very abundant in the limestones of Thunder Bay and Little Traverse Bay. Prof. Hall, the palæontologist of New York, has also afforded me many valuable suggestions, on the identification of our fossils, and the parallelism of formations. Dr. J. S. Newberry, of Ohio, who has already rendered me valuable assistance, stands ready to undertake the investigation of our fossil Flora. Capt. Meade has agreed to place at my service such maps, charts and observations of the lake survey, as may be needed in the preparation of a chapter on the Hydrography of the State; and Prof. Henry, the Secretary of the Smithsonian Institution, offers copies of such observations taken for that Institution, as may be requisite for a chapter on our meteorology.

Very many private citizens, besides the surveyors before referred to, have already communicated most valuable information on various points, which will be incorporated into my final report. Mr. James S. Lawson, of Disco, Oakland county, has furnished a description of an ancient lake terrace which is found traversing that part of the State; and I would be glad to commend this example to others who have the opportunity to make observations upon such phenomena.

Mr. A. O. Currier, of Grand Rapids, has aided me materially in arriving at a knowledge of the succession of strata penetrated in the salt borings of that place He has further provided me with a printed catalogue of the mollusca of the Grand River Valley, accompanied by a nearly complete suite of specimens

Mr. Martin Metcalf, of the same place, has likewise, in his correspondence, furnished me with important notes on the salt borings, and critical remarks on the parallelism of strata

I am indebted to Dr. DeCamp, of the same place, for a fine collection of geodes from the Grand Rapids limestone, and for fossils; and to Prof. E. Danforth for the loan of his collection of fossils from this and other States.

Dr. G. A. Lathrop, of East Saginaw, has contributed important aid in the carefully preserved series of borings taken from the first salt well at that place; in specimens and suggestions bearing upon the geology of the vicinity of Saginaw Bay; in facts and statistics illustrating the salt manufacture in the State, and by the loan of a suite of fossils

To Dr. H. C. Potter. superintendent of the salt works at East Saginaw, I am similarly indebted for important facts connected with the salt manufacture at that place.

Mr. Henry D. Post, of Holland, Ottawa county, has furnished me with observations on the outcrops of the Marshall sandstone in his vicinity.

Hon I. P. Christiancy, of Monroe, has sent the survey some interesting fossil remains from the Monroe limestone; and also statistics relative to the products of his quarries in the township of London, near Dundee. Mr. W. P. Christiancy also contributed some instructive specimens.

Thomas Crawford, Esq., of Detroit, has laid me under many obligations for polished specimens of marble from his quarry near Presque Isle; as also for some unique fossils from the same.

Mr. M. B. Hess, of East Saginaw, has supplied some desirable altitudes from the vicinity of Saginaw.

Thomas Frazer, Esq., of the Mich. C. R. R. office, in Detroit, has communiated the altitudes of the principal stations along the line of that road.

Superintendent W. K. Muir has furnished a list of altitudes of all the stations along the Detroit and Milwaukie railway, accompanied by other valuable observations.

Chief Engineer, John B. Frothingham, of Toledo, has also promised to supply me with altitudes along the Michigan Southern Railroad. It is hoped that such statistics will be further communicated by Engineers, and others, to whom they are accessible.

I am indebted to the late John Farmer for a copy of the large edition of his unequaled map of the State, and to Benjamin Fowle, Esq., for a mounted map of Hillsdale county.

Mr. John Holcroft, Superintendent of the Woodville Coal Mine, furnished me with numerous data, and other facilities, while investigating the coal formation of Jackson county.

Mr. C. E. Hovey, Superintendent of the Eagle Plaster Co., of Grand Rapids, provided me with a liberal quantity of samples of the crude and manufactured gypsum, including some ornamental vases.

Capt Malden, keeper of the light house at Thunder Bay Island, furnished me with some interesting specimens from the Huron Group. He is now engaged in a series of meteorological and tidal observations of great importance.

I am under obligations to very many of our citizens for accompanying me on my explorations, and conveying me to localities of interest, among whom, in addition to names already introduced, I may mention Hon. L. H. Parsons and Alexander McArthur, Esq., of Corunna; Benjamin O. Williams, Esq., of Owosso; Adam L. Roof, Esq., Lyons; James Scribner and J. W. Windsor, Esqs., Grand Rapids; H. S. Clubb, Grand Haven; William Walker, Jackson; Mr. W. N. Carpenter, Detroit; C. H. Whittemore, Tawas City; J. K. Lockwood and Mr. —— Melville, Alpena; the sons of Thomas Crawford, Presque Isle county; James Francis, Drummond's Island; Commissioner S. P. Mead, Sault; Langdon Hubbard, Willow Creek; J. V. Carmer, Napoleon; John Manning, London; Prof. L. R. Fisk, Lansing.

Boxes of specimens have been transmitted by C. D. Randall, Esq., Coldwater; Hon. L. H. Parsons, Corunna; J. H. Holcroft, Woodville; Wm. S. Sizer, Esq, Jackson; W. S. Brown, Grand Ledge; Dr. G. A. Lathrop, East Saginaw; Hon. I. P. Christiancy, Monroe; Francis Crawford, Esq., Detroit.

During the first season of the survey, the work was materially aided by the free passes granted to Dr. Miles and myself over the Michigan Southern, the Central and the Detroit & Milwaukee Railroads. The latter road voluntarily tendered the same appreciative acknowledgement of the importance of our labors during the present season; and I am happy here to allude to the great courtesy that has at all times been exhibited by its officers.

I should not forget to acknowledge the indebtedness of the survey to the newspaper press of the State, for numerous notices of a friendly character, calculated to awaken and increase the popular interest in the work. Among the notices which have met my eye, I am pleased to mention those of the *Michigan Argus* and *State News*, Ann Arbor; the *Commercial*, Monroe; *Patriot*, Jackson; *Clarion*, Grand Haven; *Herald*, Mackinac; *Tribune*, *Advertiser*, *Free Press* and *Farmer*, Detroit; *Enquirer*, *Eagle* and *Great Western Journal*, Grand Rapids; *Register*, Holland; *Courier*, East Saginaw; *Republican*, Lansing; *Citizen*, Flint; *Gazette*, Pontiac.

It would be impracticable to enumerate all the acts of hospitality received from our citizens; and it would be almost superfluous to say that we have been everywhere received with a welcome, and furnished with every possible facility in furtherance of our labors.

I cannot suffer the opportunity to pass without warning our citizens against lending too credulous an ear to the representations of the self-styled "geologists," itinerating amongst us. Traveling under the cloak of science, they take pains to keep out of the way of those who would detect the imposition; and instead of informing themselves truly of the geological structure of the State, prowl around the frontiers of civilization, and live upon the falsely excited hopes of a people too ready to believe that every gravel hill conceals a mine of wealth. This class of men lead their deluded followers over mounds of drift materials, they explore clay banks, they dredge the lakes, and if perchance a stray nodule of kidney ore is found, they proclaim the discovery of a mine of hæmatite; a piece of black shale turns up, and the country is rich in coal; they discover a green streak upon a fragment of limestone, and lo! copper is promised to be forthcoming in unlimited quantities. I have seen too much of this scientific quackery to allude to it with forbearance. Let the people bear it in mind, that it is not every man who styles himself a geologist who is worthy of being trusted in a geological opinion. The questions which these men attempt to decide, are the very ones most difficult for an acknowledged expert to pronounce upon. They are the last conclusions of a general and scientific survey. How can a stranger drop down in our State, without a line of knowledge of our peculiar geology, and be at once a safe adviser in important mining or quarrying enterprises. Even the man well versed in general geology may often be at fault among our formations; but most of the class of persons referred to, possess neither local nor general information. It seems unnecessary to multiply words upon the subject. Trust no "geologist" or "professor" whose credentials are not known; none who clothe their actions with an air of mystery, and hint at things which they do not plainly state, who make large pledges with small security for their performance, and have no visible means of support but what their splendid promises draw from a succession of dupes.

The act establishing the survey provides for the distribution of duplicate specimens to the University, the Agricultural College, and such other public institutions as the Governor may designate. Under these provisions, the following institutions have been designated as depositories of suites of specimens, viz.:

BY LEGISLATIVE ACT.

1. The University, *Ann Arbor.*
2. The Agricultural College, *Lansing.*

BY EXECUTIVE APPOINTMENT.

3. Mechanic's Society, *Detroit.*
4. Scientific Institute, *Flint.*
5. Lyceum of Natural History, *Grand Rapids.*
6. Young Men's Literary Association, *Kalamazoo.*
7. Young Men's Society, *Detroit.*
8. Young Men's Christian Association, Library and Reading Room, *Adrian.*
9. The Normal School, *Ypsilanti.*

Such an extended distribution of the specimens of the survey must necessarily awaken a very general interest in the energetic prosecution of the work, and the creditable elaboration of the final results. It is quite obvious, however, that this requirement multiplies the physical labors of the field geolgist, who is often called upon to carry many pounds of stones for miles, over rocky and slippery beaches, or through tangled cedar forests, in an unending conflict with musquitoes and flies, under circumstances calculated to excite commiseration. By thus increasing the amount of field work, it delays the completion of the survey. Still, there can be no doubt that the interests of the State will be best subserved by the plan proposed, even should its execution necessitate the outfit of a special party of collectors.

During the year 1859 no special attention was devoted to the Botany of the State, for the reason that the flora of the districts then under survey was already pretty well understood. All species before unobserved, all peculiarities, and some local floras were, however, noted. In the explorations of the present season, it was deemed desirable to attach a special botanical assistant to the party. Combining the observations made during the past two seasons, with notes kept by myself for several years past, I am able to present, with the aid of the University Herbarium, and Wright's Catalogue, heretofore published, a pretty complete list of the indigenous plants of the Lower Peninsula. It has not been deemed advisable to attempt to catalogue the plants of the Upper Peninsula, as the list would necessarily be defective, and it is hoped that the opportunity will be presented for completing it, next season. For local information respecting many of our plants, I am indebted to Miss Mary Clark, of Ann Arbor.

For information respecting the progress and state of the zoological survey, I would refer you to the Report of the State Zoologist.

The question is often asked when the survey will be completed. It is obvious that the answer to this question will depend entirely upon the action of the Legislature, in providing for a more or less thorough execution of the work; and upon the number of persons kept in the field. A continuance of the same provisions which have been made for the past two years, would enable us to extend the survey over the whole territory of the State, in the manner in which it has been commenced, and to furnish the final report ready for publication in three years more. It would be much better, however, to increase the number of surveying parties somewhat, with the view of effecting a more detailed examination of the unsettled portions of the State, as well as the districts which lie along the probable outcrops of those formations which possess considerable economical importance. It will not be necessary to multiply the zoological observations to the same extent as the geological. It is not necessary to identify each species of animals at every point within the limits of its general distribution; while, for the determination of the limits of the formations, this very minuteness is indispensable. Moreover, the roving habits of animals bring a large proportion of them under the notice of an observer who does not go out of his own township, while rocks must be visited in their places. It may not be amiss to state with reference to the nature of zoological field work, that a single industrious collector, employed at small compensation, would be able in one season to accumulate large stores of specimens from the remoter portions of our State. The same remark is true of botany. The elaboration of the materials thus collected must, of course, be confided to the ablest hands.

It will remain for the legislature to decide upon what scale the prosecution of the survey shall be continued. I cherish the hope, however, that provision may be made for the creditable completion of the field work, within the space of two or three years. So far as the geological work is concerned, I deem it desirable to have parties engaged, during the next season, upon the exploration of the following districts:

1st. A party upon the south shore of Lake Superior;

2d. A party upon the shores of Lake Michigan, as far as unexplored;

3d. A party in the northern portion of the Lower Peninsula.

The *personnel* required for such a prosecution of the work, besides the geologist in charge of the survey, would be as follows:

One Chemist and Mineralogist;

One Draughtsman;

Two Assistants, capable of leading parties;

Three Sub-Assistants;

Six Laborers and Boatmen.

Zoological and Botanical Collectors could be attached to the parties thus organized, with little additional expense.

I desire to close this chapter of my report with an appeal to all of our citizens to co-operate with the State Geologist in every possible way Every specimen or item of information will be thankfully received. Proprietors and managers of important enterprises, have, in some instances, greatly mistaken their true interests, in failing to furnish the data sought for, even by repeated applications. No authority is considered more reliable than a State Geological Report, on the value and extent of the mineral resources of a particular locality or district; and the interests of proprietors of mineral locations, require them to see that every evidence of the value and productiveness of their locations is placed in the possession of the State Geologist. Moreover, isolated facts or specimens calculated to throw light upon the occurrence of any rock or mineral, in any part of the State, especially those parts not yet explored, will always prove of interest, and, in some cases, may constitute critical data for deciding questions in doubt.

It will be seen, therefore, that two general classes of information are desired.

1. *Facts* calculated to contribute to our knowledge of the characters and distribution of our rocks, with their included minerals.

2. *Statistics* showing the condition of all mining enterprises and their productiveness. This class of information embraces every species of manufacture from the mineral substances of our State, as bricks, tiles, pottery, earthen ware, pipes, fire-bricks, concrete, moulding sand, glass, fluxes, land plaster, calcined plaster, alabaster ornaments, salt, its impurities, marble, quarry-stones, quick-lime, water-lime, grindstones, hones, coal, precious stones, iron, copper, lead and other metals.

When the geological department is made the common depository of all such information, the way will be opened to such a presentation to the world of our multifarious sources of wealth as will constitute the strongest possible attraction for settlement, enterprise and capital.

CHAPTER II.

DEPOSITION, DISTURBANCE AND DENUDATION OF STRATA—GENERAL PHYSICAL STRUCTURE OF THE NORTHWEST.

The geological series in our State is very complete from the horizon of the oldest known rocks, to the top of the Carboniferous System. From this point to the Glacial Drift, the formations observed in other parts of the county are, as far as investigations have extended, entirely wanting. All that portion of the Michigan series lying above the Niagara Group, is found within the limits of the Lower Peninsula; while the Niagara Group and all rocks below, are confined to the Upper Peninsula and the islands at the head of Lake Huron.

The rocks of the Upper Peninsula not having as yet come under the observation of the present survey, it is not deemed necessary to refer to them at the present time, any further than to show their connection with the geology of the contiguous districts.

In order to convey a clear idea of the superposition and lines of outcrop of our different rocks, it will be desirable to offer a few words on the general conformation of the strata of the Northwest. Although the stratified rocks of the country succeed each other in regular ascending order, it must not be supposed that these strata always occupy a horizontal position, that they are necessarily continuous between distant points, or that any given stratum is always actually overlain by those strata which belong higher in the series. The sediments from which these rocks were formed, were seldom deposited in perfectly horizontal sea bottoms, but to facilitate our explanation, we may suppose that they were. We will suppose, also, that one series of sediments was deposited upon another for the space of many ages, and forming a thickness of several thou-

6

sand feet. We have thus the materials for several geological formations, each with its own mineral characters, and embracing the organic debris which characterized its own age. Through some appropriate agency these sediments become solidified. But at length some movements begin to be experienced by the solid crust of the earth, and our horizontal strata begin to be elevated in one place and depressed in another. Here is a dome shaped bulge, and there is a long ridge, rising in some of its parts above the surface of the sea. Successive disturbances increase the inequalities, and at length our level sea-floor presents all the irregularities of a carpet carelessly thrown down. By degrees the general uplift of the sea bottom has made an extensive addition to the continent.

Thus far we suppose each successive layer of rock to be continuous over every ridge and through every valley. But now we must consider the effect of *denuding* forces—those forces which move over the surface, and plane down the inequalities. Whether these results are attributable to the action of the atmosphere, frost, glaciers, powerful currents of an invading sea, floating icebergs, or to all of these agencies combined, or in succession, cannot here be considered. It is sufficient to know that such forces have acted, and that all the original elevations have been more or less worn down, and the rubbish produced strewn over the general surface, tending still further to obliterate its unevenness. Consider what would be the effect of paring off the summits of the ridges and domes of upraised strata. The uppermost layer would be sliced through, and the second in order would come in sight. Then the wearing would continue till the second layer would be cut through, and the third would appear. So, in some cases, the denudation has continued, till thousands of feet of strata have been pared off, and the underlying granite has been exposed; and then this has been planed down some hundreds of feet. Glance now at the cut edges of the strata. The lowest rock reached will be found in the center of the dome, or along the central axis of the ridge. If it is a dome, the overlying strata dip in all directions from the center. If it is a ridge, they dip to the right and left of the axis. This ridge may have been planed down to the general level of the country. If this is the case, we shall then, in passing from the central line either to the right or left, pass continually from lower to higher rocks, withour changing our elevation. We ascend stratigraphically, but not topographically.

This ridge may not pursue a straight course. It may finally bend round, and proceed in a direction parallel with itself. It is obvious then, that the strata between the two portions or branches of the ridge, form trough-shaped depressions. In many cases *all* the edges of the over-lying strata are turned up, and they rest in a dish shaped depression. When the irregularity of the original elevations is considered, it is obvious that the outcropping edge of any stratum, when traced along over the surface of the earth may pursue a very tortuous course, or *strike.* It is also obvious that the width of the stratum at the surface will be more, if the surface cuts it very obliquely, less, if the surface cuts it nearly at right angles. This depends, in other words, upon the *amount* of the dip; so that a thick formation, by being nearly vertical, may occupy a very narrow belt of country; while a thin one, by being nearly horizontal, may occupy a belt several miles in width.

All this is familiarly illustrated by the lines of the "grain" of a smoothly planed board, especially if slightly gnarly or knotty. The knots may represent the granite, while the layers of wood surrounding it—here apparently thin, because cut nearly at right angles, there spreading out, bocause cut more obliquely, here running in a straight line, and there tracing a zigzag path—may represent the layers of rock, occupying a geological position above the granite.

These explanatory observations are here admitted, in the hope of obviating some difficulties almost always experienced by persons unversed in geology, in forming general conceptions of the geological structure of a particular region.

The wide interval between the Alleghany and the Rocky

mountains was once an ocean bed, over which were strewn the various sediments that have formed the groups of rocks, which stretch with more or less regularity from one end of this area to the other. Geological agencies have left this ocean floor in an undulating position; and subsequent denudation of the higher points, has worn many holes through the upper layers of rock, where they have been pushed up into exposed attitudes. The city of Cincinnati stands upon a dome of older strata, which have been uncovered by the planing off of the higher beds. The strata dip in every direction from this vicinity. Toward the north, however, the dip is least, and something of a ridge extends towards the common corner of Ohio, Indiana and Michigan. It bifurcates, however, before reaching that point, and the east branch runs up to Monroe county, crosses Lake Erie and subsides in Canada West; while the west branch passes across northern Indiana and Illinois, to the head of Lake Michigan, and thence north-westward.

A ridge extends through Canada, along a line nearly parallel with the St. Lawrence, to the region north of Lake Ontario, and thence trends north-west around the northern shores of Lakes Huron and Superior. The rocks around the shores of Lake Huron dip south-west and south, away from this ancient axis of elevation.

It appears, therefore, that the Lower Peninsula of Michigan is surrounded on all sides by ancient axes of elevation; and even if the surrounding regions do not in all cases actually occupy a higher level, we must expect to find the strata dipping from all sides toward the centre. Each rocky stratum of the Lower Peninsula is, therefore, dish shaped. All together, they form a nest of dishes. The highest strata are near the centre of the peninsula; and passing from this point in any direction, we travel successively over the out-cropping edges of older and older strata. The irregularities in the shape of these dishes, will be pointed out in the sequel.

The southern part of the Upper Peninsula is covered by the lower members of the southward dipping series, whose upper members are found in the Lower Peninsula, and whose axis of elevation lies north of the great lakes. At Marquette, Keewenaw Point, the Porcupine Mountains, and other localities, however, we find accessory axis of elevation, giving rise to dips in various directions, which will be explained on some future occasion. Lake Superior occupies a valley between the elevations on the north and south shores, while the other lakes rest in troughs, which have been excavated nearly along the outcropping edges of some of the softer formations. On the south, a basin similar to that of lower Michigan, occupies the southern part of Illinois; while, passing east from Sandusky, in Ohio, we begin to step over the north-western limits of another one, which reaches to the Alleghanies, and in the other direction stretches from New York to Alabama. Still further west, another basin rests, with its northern border in Iowa, and its southern in Missouri.

A knowledge of these great undulations in the wide-spread strata of the north-west, and of the effects of denudation of the crests of the elevations, will aid materially, in connection with the descriptions which follow, in giving definite ideas of the geological structure underlying any particular portion of our State.

CHAPTER III.

GENERAL SKETCH OF THE GEOLOGY OF MICHIGAN, AND ITS CONNECTION WITH SURROUNDING DISTRICTS.

The rocks which constitute the solid crust of our earth may be arranged into great groups according to the following plan:

STRATIFIED.

Fossiliferous.

Azoic, or unfossiliferous.

UNSTRATIFIED.

Volcanic, as lava, trap, &c.

Plutonic, or Granitic, as granite, syenite, &c.

Geologically speaking, the Fossiliferous strata are higher than the Azoic, while the place of the Plutonic is generally below the Azoic; and the relative antiquity of these three classes of rocks is represented by this order of superposition. The volcanic rocks have burst up through the other rocks at various periods, and the same is to some extent true of the Plutonic—some new granites appearing to have been formed since the granitic substratum of the Azoic rocks was formed. The Upper Peninsula furnishes us with abundant examples of all these classes of rocks. After devoting a few words to the unstratified rocks, we shall proceed to speak of the stratified, as nearly as possible, in chronological order, beginning with the oldest.

I.—PLUTONIC GROUP.

A belt of granitic rocks comes down from the northwest into northern Wisconsin, and encroaches a few miles over the Michigan boundary line between Montreal river and Lac Vieux Desert. At the surface this is separated by a belt of Azoic rocks from another mass of granite, which is probably a continuation of the first, and which begins near the head waters of the Sturgeon river, and extends east, gradually widening, until it occupies the region a few miles back from the lake coast, all the way from the Huron river to Presque Isle, at which two points it abuts upon the coast, reappearing again in the Huron islands on the west, and Granite Island on the east. Another granitic boss rises up in the district south of the Iron Region, and covers about twelve townships, and still others, on a small scale, are found east of the mouth of the Machigamig river.

The rock throughout these exposures is seldom a true granite, being composed mostly of feldspar and quartz, with occasional intermixtures of mica in small quantity. Hornblende sometimes replaces the mica, and the rock becomes syenite. The plutonic rocks on the south shore of Lake Superior, appear to have been upheaved after or towards the close of the Azoic period.

II.—VOLCANIC GROUP.

A range of volcanic rocks extends from the extremity of Keweenaw Point to Montreal river, running nearly parallel with the lake coast, and having a width varying from two to eight miles. About twelve miles east of Montreal river the belt suddenly widens to about fifteen miles, sending a spur off on the south side toward the southern extremity of Agogebic lake. Another spur sets off north to the Porcupine Mountains. To the east of Portage lake this belt is in reality two belts—the "northern" one consisting of interstratified masses of amygdaloid, conglomerate and coarse sandstone; the "southern," or "Bohemian" range being a mass of crystalline trap. About a mile north of the northern range, another narrow belt curves round parallel with the coast from a point opposite Manitou Island, to the eastern point of Sand Bay. The belt called the Northern Range contains the larger number of copper locations. These rocks were erupted during the period of the Lake Superior Sandstone. A contemporaneous range forms the basis of Isle Royal.*

III.—THE AZOIC SYSTEM.

An immense thickness of unfossiliferous strata is interposed between the crystalline rocks just referred to, and the Lake Superior sandstone. These, in the Upper Peninsula, commence at

* For information concerning the rocks of Lake Superior, see Foster and Whitney's Report.

and near the mouth of Chocolate river, and extend westward to join another belt beginning a few miles south of Huron river. The first belt in the neighborhood of the Machigamig river, suddenly expands towards the south, so that on the State boundary the Azoic belt stretches from beyond Lac Vieux Desert to Chippewa Island, in the Menomonee river. It extends thence westward through Wisconsin and to the sources of the Mississippi. The rocks of this system consist in Michigan of talcose, chloritic and silicious slates, quartz, and beds of marble. The silicious slate, becomes, near Marquette, a novaculite, from which hones have been manufactured. In this system are found the specular and magnetic iron ores of Lake Superior, as well as of Pilot Knob, and perhaps the Iron Mountain, in Missouri, the Adirondacks of New York, and other localities. This series of rocks attains an enormous thickness on the northern shores of Lakes Superior and Huron; and Sir Wm. Logan, the Director of the Canadian Geological Survey, has decided that they constitute two great systems, unconformable with each other, the upper of which he styles the *Huronian* series and the lower the *Laurentian.** The Bruce, Wellington, and neighboring mines, are located in these rocks, and are worked for the *ores* of copper; while the Lake Superior mines are located in veins which belong to the age of the trap, and are worked for *native* copper.

IV.—FOSSILIFEROUS STRATA.

I.—LOWER SILURIAN SYSTEM.

1.—*Lake Superior Sandstone.*

The reddish, yellowish, grayish or mottled sandstone, found along the south shore of Lake Superior has, by different writers, been assigned to different geological periods; but the weight of authority is decidedly in favor of placing it at the base of the Palæozoic series, and on the horizon of the Potsdam Sandstone of New York. Further examinations will undoubt-

*Report 1852-3, p. 8; 1856, p. 171.

edly result in the discovery of data which will settle beyond cavil this long mooted question.

The examinations of the past season have found this sandstone in place at the Falls of the St. Mary's river, where it has a measured thickness of at least 18 feet. It is here thin bedded, moderately coherent, reddish and blotched with gray, or grayish blotched with red. It presents evidence of having been deposited on an uneven sea bottom, and in shallow water. We find local undulations, and very distinct ripple marks. On some of the surfaces are obscure traces of *Algæ*. On some specimens from the Montreal river, not less than three species of fossil plants have been discovered; "sun cracks" are also frequent. This sandstone is believed to underlie the whole of Sugar Island, and the northern extremity of Sailor Encampment Island. On the Canada shore, opposite the Neebish Rapids, an altered sandstone is found, which apparently belongs to the same formation. It is of a light gray color blotched with reddish-purple spots, and having a rapid dip S. 55° W. It is intersected by nearly vertical divisional planes, running at right angles to the dip. Near the northwestern extremity of St. Joseph's Island, a quartzose sandstone appears, striped and banded with red along lines which appear to mark the original planes of stratification. A little further east, rock is again seen resembling that at the Neebish Rapids, and having a dip of 20° toward S. 55° W. On the south-east shore of the bay which indents the northern extremity of St. Joseph's island, a jaspery conglomeratic sandstone is seen, rising in small rounded knobs, possessing a general reddish color, and being destitute of obvious stratification. The small islands at the southern angle of the channel which separates Campement d'Ours from St. Joseph's Island, are formed by the same rock. It will hereafter be seen that these quartzose and conglomeratic sandstones occur in close proximity to fossiliferous limestones. Quartz rock is next seen on Sulphur Island, north of Drummond's. It is slightly clouded with reddish spots, and occurs in beds from three to six feet thick, with shaly partings. It immediately underlies a limestone containing fossils in a perfect state of preservation. A conglomerate also occurs here, made up of rounded masses of quartz, ranging from the size of a pea to boulders many tons in weight, all cemented together by a silicious limestone, not altered, but appearing as if deposited amongst the interstices and open spaces of a pile of stones and gravel.

The solid quartzose character of the rock on St. Joseph's and Sulphur Islands, so unlike the conglomerate and altered sandstone of Lake Superior, seems to suggest the idea of its being *azoic*, and it is so colored on Foster and Whitney's map, where, nevertheless, it is made to appear like a prolongation of the Potsdam sandstone of Sugar Island. The gradual transition, however, from the unaltered sandstone of the Sault, to the altered sandstone of Neebish Rapids and the extremity of St. Joseph's Island, the quartzose sandstone and jaspery conglomerate of the shore west of Campement d'Ours, and the quartz and conglomerate of Sulphur Island, favors the idea of the equivalency of the sandstone and quartzose rocks. The superposition of fossiliferous limestone, at Sulphur Island (probably the Chazy limestone) immediately upon the quartzite, favors the same inference, inasmuch as there is no probabiilty that the sandstone would not be interposed at this place between the Chazy and the azoic rocks. Moreover, the influence of the igneous disturbances which have taken place at the Bruce mines and along the Canadian shore but a few miles distant, furnish sufficient cause for the alteration suggested. The Canadian geologists have frequently recognized the Potsdam sandstone in a similar condition.

2.—*Calciferous Sandstone.*

Though this formation, as just stated, is not recognized to the east and south of St. Mary's Falls, it is thought best to embrace it in the enumeration, since it is represented as playing an important part in the geology of the country west of St. Mary's river.

3.—*Trenton Group.*

The gray silicious limestone, seen resting on quartz on Sulphur Island, north of Drummond's, is regarded as the lowest fossiliferous limestone within the limits which have come under observation. On the south and south east sides of Copper Bay, in Montreal Channel, is observed a series of limestones supposed to belong but a short distance higher up. The following fossils have been recognized from the extensive collections made along this shore. I have not the time at present to classify them stratigraphically; nor would such particularity comport with the scope of the present report. They are referred to their localities:

At 758, *Rhynchonella plena.*

At 760, in fragments on the beach, *Subulites elongatus, Cypricardites ventricosus, Murchisonia bicincta.*

At 762 A.=d. in the Synoptical Table, next chapter, *Rhynchonella plena, Rhynchonella altilis, Strophomena (n. sp.)*

At 763=762 F.=upper part of 770 D., *Schizocrinus, Leptæna sublenta, Subulites elongatus, Cypricardites ventricosus, Pluroto-maria subconica, Rhynchonella plena, Asaphus gigas.*

At 764=762 B., *Strophomena camerata.*

At 766, (not in place) *Rhynchonella plena, Tetradium cellulosum.*

At 769, the north-eastern extremity of the headland on the south-east side of Copper Bay, *Leperditia fabulites, Leptæna sub-tenta, Dalmannites callicephalus, Ambonychia amygdalina, Strophomena plicifera, Cypricardites (sp?), Asaphus gigas, Orthoceras anellum (?) Pleurotomaria subconica, Illænus, Subulites n. sp, Cypricardites ventricosa, Murchisonia, (sp ?) Orthis bellirugosa, O. trisenaria.*

At 771, *Strophomena filitexta, Receptaculites*

At 785, *Rhynchonella plena, Orthis* (resembling *O. pectinella,* but distinct).

At 786, *Streptelasma corniculum.**

*The region from which the foregoing fossils were obtained, is colored on Foster and Whitney's map, as lying along the northern margin of the Calciferous Sandstone belt. For identifications of species I am under great obligations to Prof. Hall.

The Trenton Group of rocks forms a belt about four miles wide, extending west-northwest across St. Joseph's Island, re-appearing in the high bluffs opposite Little Sailor Encampment Island, and extending thence across the middle of Great Sailor Encampment Island. From here it stretches west in a gradually widening belt, which, bending round to the south-west, lies with its southern border on the west shore of Little Bay de Noquet and Green Bay, whence it continues across Wisconsin into northern Illinois.

4.—*Hudson River Group.*

On the north side of Drummond's Island are found some highly argillaceous limestones abounding in the fossils characteristic of the Hudson River Group of New York. These are first seen about three miles west of Pirate Harbor, and extend thence around the coast to the point of land north of the bay which indents the north-west side of the island, thus occupying a belt about three miles wide. A large proportion of the fossils seen are Bryozoa, which have not yet been studied. *Chetætes lycoperdon* is exceedingly abundant. *Favistella stellata* occurs in prodigious masses and great numbers, (at 781, 786, 788.) At 781 (A) is an *Ambonychia* not yet identified. The argillaceous strata are about fifteen feet thick, and underlain by a bluish gray, subcrystalline limestone, of which three feet were observed.

This group forms a belt about four miles wide across St. Joseph's Island, a little south of the middle, then, intercepting the southern extremity of Great Sailor Encampment Island, stretches westward along the south side of the region covered by the Trenton Group, and occupies the space between Big and Little Bays de Noquet. Passing under the whole length of Green Bay, it reappears at the southern extremity, and continues in the direction of Winnebago and Horicon lakes, in Wisconsin.

All round the circuit which is thus traced, the dip of the formation carries it under the lower peninsula of Michigan. It does not emerge on the southern side of the peninsula, being overlain by the four groups next described, but dips down again beneath the carboniferous basin of Ohio, on the one hand, and of Indiana on the other. At Cincinnati is another swell, from the summit of which the overlying formations have been denuded, and here the Hudson River Group again appears. Like most of the other groups of the Palæozoic System, it has throughout the northern and north-western States, a very great geographical development.

II.—UPPER SILURIAN SYSTEM.

5.—*Clinton Group.*

At the eastern extremity of Drummond's Island, the lower 32 feet of Dickinson's quarry constitute the upper portion of the Clinton Group of New York. . It is an argillo-calcareous limestone, fine grained and very evenly bedded, in layers from two to three feet thick, having a very gradual dip toward the south. In color it is nearly white, some layers having an ashen hue. The rock presents to the eye every appearance of a most beautiful and desirable building stone, remarkably easy of access and eligibly situated for quarrying. In November, 1859, the company organized for working the quarry got out a large quantity of fine blocks for building purposes. Severely cold weather arrested their operations, and on the return of Spring, the fine blocks quarried out were found considerably shattered, apparently by the action of the frost. This effect was undoubtedly due to the sudden freezing of the stone while yet containing a large amount of quarry water. Whether a rock containing so large a per centage of argillaceous matter would not, under any circumstances, prove too absorbent and retentive of moisture, to stand in exposed situations in our severe climate, remains yet to be ascertained; but I have some hope, that if quarried in early summer, and left to dry before the approach of frost, it might be found durable.

As a lithographic stone, whatever its qualities for building purposes, I believe some of the layers will answer well, when polished ; the surface, to the naked eye, is quite free from imperfections, and under a glass some portions are so homogeneous as to seem made from an earthy impalpable powder. I have not had the opportunity, however, to submit any samples to the inspection of a competent lithographer, which alone would decide the value of the rock for this purpose.

Rocks lower in the group are seen outcropping successively along the shore of the island, from Dickinson's quarry to Pirate Harbor, which, as before stated, is not more than three miles, nearly along the strike of the formation, from the first appearance of rocks of the Hudson River Group. The formation reappears on the northwestern side of the island at Brown's and Seaman's quarries (790 and 796). An experimental quarry of the ship canal company was opened in this vicinity in the upper part of the group (792), but was subsequently abandoned. The same rocks are seen at numerous points as far south as the neighborhood of the old British Fort, the northern extremity of the point of land at the west end of the island being of the Clinton Group.

The rocks of this group contain few fossils, but among our collections I recognize the *Avicula, Murchisonia* and *Cytherina (Leperditia?)*, referred to by Prof. Hall. The latter, particularly, is characteristic at all the localities, and throughout the whole vertical range of the formation. At Dickinson's quarry, some arenaceous layers are seen above the Clinton rocks from four to six inches thick, somewhat blotched with red, and strongly ripple-marked. The resemblance to the Medina sandstone is so strong that one expects next moment to find *Lingula cuneata* in it, but careful search has revealed no organic remains.

This group cuts across the southern part of St. Joseph's Island, and passes on in the direction of the southern shore of Munnusco Bay.

6.—*Niagara Group.*

The principal part of the promontory known as Marblehead at the eastern extremity of Drummond's Island, is composed of the Niagara limestone, so called by the New York geologists, from its occurrence on the Niagara river. It reaches here an elevation of nearly 100 feet above the lake, and dipping southward sinks beneath the water on the south shore of the island. This assemblage of strata embraces a band five feet thick of highly arenaceous limestone, at bottom, overlain by seven feet of a hard, gray crystalline limestone, which furnishes an excellent quality of quicklime. This is overlain by forty-five feet of a rough, crystalline, geodiferous limestone, followed upward by eight feet of broken thin-bedded limestone, and six feet of rough vesicular limestone. The white, massive, marble-like, magnesian limestone, twenty feet thick, occupying the south shore of the island, is still higher; and the series is completed by about six feet of thin bedded brown limestone, abounding in *Favosites niagarensis*, *Halysites escharoides*, *Heliolites spinipora*, &c. The thicker masses are eminently characterized by *Pentameri*, while not one has been found in the Clinton Group. The total observed and measured thickness of these rocks does not exceed one hundred feet, and it is doubtful whether the dip of the strata across Drummond's Island would give them a calculated thickness much greater. The rocks which emerge from the water on the south side, preserve a gentle and pretty uniform rise to the top of the escarpment at Marblehead, and west of there. Only the uppermost, thin-bedded layers seen on the south shore, are wanting at Marblehead.

The economical qualities of this limestone, so far as I am aware, have not been reliably tested. The large per centage of carbonate of magnesia contained in the heavier beds, renders them a pretty well characterized *dolomite*. According to the researches of Vicat, this proportion of carbonate of magnesia, mixed with about 40 parts of carbonate of lime, possesses hydraulic properties; and only a few hundredths of clay are required to be added, to produce the strongest hydraulic cement. It is not at all unlikely that somewhere upon the shores of Drummond's Island a good hydraulic limestone may be found compounded by the hand of nature.

At several points on the south shore of the island, the thick strata above the Pentamerus beds, appear well calculated for architectural uses. The rock is highly crystalline, hard and white, with occasional stripes and blotches of a rose color, and can be conveniently procured in blocks of any required size. It is not at all unlikely that quarries may be opened which will furnish a stone sufficiently homogeneous to be used for ornamental purposes. For rough, substantial masonry, there is no rock in our State which is more worthy of attention; and when once developed, there will be no building stone of equal excellence half as accessible to our people.

This group of rocks occupies the southern portion of the Manitoulin chain of islands to the east and south east of Drummond's, underlying the peninsula between Georgian Bay and Lake Huron, and stretching thence to Hamilton, in Canada West, crossing the Niagara river between Grand Island and Lake Ontario, and forming at Lockport, in New York, the quarry stone which has been sent a thousand miles to build the steps at the St. Mary's Ship Canal.

Toward the west the Niagara Group occupies the whole shore as far as Point Detour of Lake Michigan, except the promontory, west of Mackinac. Continuing south-west, it forms the Potawotomie Islands, and the peninsula between Green Bay and Lake Michigan, the coast of which it does not leave until it reaches the neighborhood of Evanston, near Chicago.

7.—*Onondaga Salt Group.*

On the east side of Little St. Martin's Island, north of Mackinac, is seen at the surface of the water a mass of gypseous mottled clay, constituting the lowest beds of the Onondaga Salt Group of New York. On the main land west of Mackinac the clays again appear, and in the vicinity of Little Pt. aux Chene they are seen inclosing numerous masses of aggregated crystals of brown and gray gypsum. From the latter locality several ship loads were at one time sent off, but the business was interrupted by the death of one of the proprietors, and has not since been resumed.

At a higher level, we find at the base of Mackinac, Round and Bois Blanc Islands, as well as at Sitting Rabbit on the main land west, a fine, ash colored argillaceous limestone, containing abundant acicular crystals, and becoming in the lower part banded with darker streaks of aluminous matter, and resembling the water limestone of this group in New York. Above this, at the west end of Bois Blanc Island, are found three feet of calcareous clay or marl; while still higher and immediately underlying the rocks of the next group, occurs at all the above localities, a fine-grained, brown limestone. No fossils have been discovered in this group, in the northern part of the State.

From the region just referred to, this belt of rocks passes under the bed of Lake Huron, reappearing on the Canada shore between the river Au Sauble and Douglass Point. It thence extends to Galt, in Canada West, and crosses the Niagara river south of Grand Island. On the west, it passes in a similar manner under the bed of Lake Michigan, and barely makes an outcrop in the vicinity of Milwaukee, whence it has not been certainly distinguished from the associated limestones of the Clinton, Niagara and Helderberg groups, the entire assemblage being commonly known as the "Cliff Limestone."

No other outcrop of rocks of this group has heretofore been known in our State. I have now, however, to announce the existence of the Onondaga Salt Group in Monroe county, in the south-eastern corner of Michigan. My attention was first attracted by the peculiar character of the limestones at Montgomery's quarry, in the south part of the township of Ida. At this place I found the characteristic acicular crystals in great abundance, in a light, thin-bedded, fine-grained, argillaceous limestone; and discovered also, some beds of the brownish banded argillaceous rock forming the water-limestone of the group. At this place occur the only fossils yet detected in the group in this State. They consist of a turrited gasteropod (*Laxonema Boydii?*) and an obscure Cyathophylloid coral.

The group was again recognized at the head of Ottawa Lake, in the south western part of the county, and again at numerous points in the bed of Otter Creek, in the eastern part of the county. The deepest of the Plumb Creek quarries, two miles south of Monroe, have penetrated the same formation and revealed marked and satisfactory characters.

Since making the above observations, I have been informed of the discovery of gypsum at Sylvania, in Ohio, just beyond the State line, and am led to regard this as confirmatory evidence of the distinct existence of this group in the southeastern part of our State. It might not be too much to allege that the gypsum exported from Sandusky, probably holds a position in the same geological horizon.

The economical importance of the Onondaga Salt Group of rocks is very great. It is the source of all the salt and gypsum of the State of New York, and supplies at Galt, in Canada West, a beatiful stone for building purposes. In our own State it has been already shown to contain gypsum in workable quantities on the shores of the Upper Peninsula, near Little Pt. au Chene. The occurrence of gypsum at Sandusky and Sylvania, in Ohio, justifies the search for it in Monroe county. The localities most favorable for exploration are those already mentioned, viz.: the deepest excavations at Montgomery's quarry, the Plumb Creek quarries, those at the head of Ottawa Lake, and the gorges of Otter Creek.

Some indications likewise exist, of the saliferous character of this formation, in Michigan. Occasional salt springs occur in Monroe county, far beyond the outcrop of the saliferous sandstones of the center of the State. The most noteworthy of these is 4½ miles south of the Raisinville quarries, in the township of Ida. An Artesian well sunk at Detroit in 1829-30, after passing through 130 feet of unsolidified materials, and 120 feet of compact limestone, passed 2 feet of gypsum containing salt. On the opposite side of the State, according to information furnished by Dr. Miles, is a strong and copious salt spring, located upon Harbor Island in the west arm of Grand Traverse Bay. This is now overflowed by the waters of the lake, but tradition says that the Indians formerly manufactured salt at this place, when the water was several feet lower. It appers quite possible, therefore, that borings which should penetrate this group of rocks might be rewarded by a profitable supply of brine.

One other suggestion may be made in connection with the economy of this group. The brown and banded argillaceous limestone, which, in Monroe county, generally occurs in the deeper parts of the quarries, may, on trial, be found to produce a valuable water lime. The trial, if never made, should, by all means, be undertaken. Even should this experiment fail, the hydraulic character imparted to the quicklime manufactured from this rock, or from rock with which this is mixed, must add materially to the cementing properties of the lime, *provided* it is used with reference to the peculiar nature of hydraulic cements.

III.—DEVONIAN SYSTEM.

8.—*Upper Helderberg Group.*

In the lower part of the cliffs known as Chimney Rock and Lover's Leap, on the west side of Mackinac Island, is seen a cherty and agatiferous conglomerate, irregularly disposed, but pretty persistent. On the main land west, close to the water's edge, and beneath the brecciated mass, presently referred to, is found a better characterized conglomerate, a few feet in thickness. These beds, occupying the place of the Oriskany Sandstone of New York, and corresponding to it in lithological characters, as seen at some of its exposures, may not improbably be regarded as representing that formation. The uncertainty of the identification, however, prevents me from giving it a distinct place in the enumeration of our strata.

Above this curious conglomerate, rises one of the most remarkable masses of rock to be seen in this or any State.

The well characterized limestones of the Upper Helderberg Group, to the thickness of 250 feet, exist in a confusedly brecciated condition. The individual fragments of the mass are angular and seem to have been but little moved from their original places. It appears as if the whole formation had been shattered by sudden vibrations and unequal uplifts, and afterwards a thin calcareous mud poured over the broken mass, percolating through all the interstices, and re-cementing the fragments.

This is the general physical character of the mass; but in many places the original lines of stratification can be traced, and individual layers of the formation can be seen dipping at various angles and in all directions, sometimes exhibiting abrupt flexures, and not unfrequently a complete downthrow of 15 or 20 feet. These phenomena were particularly noticed at the cliff known as Robinson's Folly.

In the highest part of the island, back of Old Fort Holmes, the formation is much less brecciated, and exhibits an oolitic character, as first observed in the township of Bedford, in Monroe county. The principal part of Round and Bois Blanc Islands is composed of the brecciated mass. It forms the promontory west of Mackinac, which, on the north side, sinks abruptly to the low outcrop of the Onondaga Salt Group, stretching across from the Hare's Back to Little Pt. au Chene. It is seen again in the vicinity of Old Mackinac, but it evidently diminishes in thickness toward the south.

The elevated limestone region constituting the northern portion of the peninsula, consists of the higher members of the Upper Helderberg Group, which gradually subsides toward the south, and in the southern part of Cheboygan county, as nearly as can be judged, sinks beneath the shaly limestones of the Hamilton Group. The strike of the formation determines the trend of the coast of Lake Huron, although the limestone barriers to the lake are generally, at the present day, situated some distance back from the immediate shore. A few miles north-west of Adam's Point, at Crawford's marble quarry, the higher members of the series abut upon the shore in a cliff about seventy-five feet high. At the base we find four feet of brown calcareous sandstone which is assumed to be next in order above the oolitic beds of Mackinac Island. From this point, the outcrop of the formation is traced in a ridge passing between Grand and Long Lakes, in Presque Isle county, and abutting upon the shore again at a point nearly opposite Middle Island. This island is made up of fragments of the limestone. Gradually subsiding toward the south, the formation at Thunder Bay Island rises barely to the surface of the water. On the east side of the island, in the vicinity of the light-house, it is seen forming vertical cliffs beneath the surface of the water. In calm weather, upon a sunny day, the view of these subaqueous precipices is truly impressive. Dark gorges, gloomy caverns and perpendicular walls are seen dimly lit by the diminishing light, until darkness cuts off the view, and the plummet feels its way to the depth of ninety feet, amongst the shadows of the ruins of an ancient ocean stream. Passing hence under the bed of the lake, the formation emerges on the Canadian shore, between Douglass Point and Benson's Creek. It passes thence in a broad belt to the shore of Lake Erie, which it occupies between Buffalo and Long Point. Dipping toward the south-west beneath a trough of newer rocks, it appears again upon the northern shore of the lake between Point aux Pins and the Detroit river, and passing into south-eastern Michigan, it arches over, forming the anticlinal axis whose denudation has uncovered the Onondaga Salt Group. From this axis it dips north, south-east and south-west, passing beneath three distinct coal basins.

At the exposures of this group of rocks in the southern part of the State, we find its thickness very considerably diminished. The conglomerate, supposed to represent the Oriskany sandstone, has not been recognized. The thick brecciated mass is not distinctly identifiable, though at Pt. aux Peaux and Stony Pt., the formation is much broken up. Still the palæontological characters of the rock seem rather to ally it with that part of the formation seen at Thunder Bay Island. The oolitic portion seen at the summit of Mackinac Island is recognized at several points in Monroe county, while the arenaceous strata of Crawford's quarry are repeated in a beautiful white sand, derived from the disintegration of the rock in Raisinville, 8 miles from Monroe. The whole thickness of the formation in Monroe county cannot be over 50 or 60 feet from the oolitic beds to the Onondaga Salt Group, while at Mackinac the same strata attain a thickness of 275 feet.

To the west of Mackinac, the Helderberg limestones are found underlying the numerous islands near the foot of Lake Michigan, and forming the highlands seen a few miles back from the coast of the Peninsula, as far as Little Traverse Bay. At the head of this bay, they are seen forming cliffs along the shore. The highest beds are thick, light, argillo-calcareous, regularly stratified, abounding in Brachiopods, geodes and long cylindrical cavities. At some points these beds are made up of a large dome shaped coral, similar to those seen at Thunder Bay Island. A calcareo-argillaceous, shaly layer, of a dark gray color, one or two feet thick, separates these upper beds from a pale buff, argillo-calcareous, thick bedded, fissile mass, 4 feet thick, which is underlain by 3½ feet of a light dingy gray argillo-calcareous, porous, geodiferous mass, breaking with a very uneven fracture. Still lower we find 6 feet of light argillaceous, fine grained limestone, resembling that of the Clinton Group. We next come to a light buff limestone, much shattered, destitute of fossils, 6 feet thick, apparently representing the brecciated mass about Mackinac. Finally, at the lowest points, is seen a light buff limestome, banded with argillaceous matter, and resembling the highest beds of the Onondaga Salt Group.

The Helderberg limestones of Michigan are well stocked with fossil remains, which are found not only in place, but scattered with the drift to all parts of the State Probably three-fifths of all the fossils picked up from the surface of the Lower Peninsula—except in the immediate vicinity of the outcrop of other fossiliferous strata—belong to this group; while more than another fifth belong to the Hamilton Group. But little has yet been done toward the identification of the numerous species, in consequence of the long expected, but long delayed, appearance of Prof. Hall's third volume on the Palæontology of New York. The highest members of the formation in Monroe county, contain numerous ichthyodorulites and other traces of fishes, the most perfect of which have been furnished by Judge Christiancy, from his quarry near Dundee. A finely preserved spine from this locality, exhibits the generic characters of Newberry's *Machœracanthus** except that it is solid throughout. I have also a traditional account of a pair of powerfully armed fish jaws. The same quarry contains an abundance of beautifully preserved *Tentaculites*, showing the telescopic structure of the shell; a large encrinital stem, and a *Gomphoceras* (n. sp.), which is found again in the highest beds of the formation at Crawford's quarry, beyond Presque Isle. A little lower down, in the borders of the oolitic beds, we find a *Rhynchonella* (n. sp.). At Stony Pt. and Pt. aux Peaux, the formation is much shattered, and embraces large concretionary masses several feet in diameter, which easily separate in concentric layers. A similar structure was afterwards seen at Thunder Bay Island, forming domes twelve and a-half feet in diameter, rising up through the rocky floor of the island. Here, however, a distinct coralline structure was discovered, which has led to the conviction that the structure at Stony Point, is also organic. Numerous trilobites occur in the rocks at Monguagon, in Wayne county, among which *Phacops bufo* is conspicuous. Two or three species of *Euomphalus* were seen at Middle Island, and a very large Euomphaloid shell six or eight inches across, has been obtained from the west end of Lake Erie. From Mackinac, besides *Phacops bufo*, *Proetus* (sp?) and the other forms noticed by Prof. Hall,† I have detected only a Cyathophylloid coral. From Little Traverse Bay, I have *Spiri-*

*"Fossil Fishes from the Devonian Rocks of Ohio," in *Bulletin of the National Institute*, Jan. 26, 1857.

†Foster & Whitney's Report, Vol. II, p. 160.

fer gregaria, Merista, Cyrtia, (n. sp.) *Acervularia Davidsoni* (from the limestones separating the Helderberg and Hamilton Groups,) and numerous other fossils. From other parts of the State, this group has furnished a *Bellerophon,* (n. sp.) *Spirifer acuminatus, (cultrijugatus,) Syringopora, Chonetes, Productus, Atrypa reticularis, Strophomena rugosa, Spirifer duodenaria, Strophomena hemispherica, Atrypa (n. sp.), Spirifer* (peculiar for plication in mesial sinus) *Strophodonta (n. sp.) Meristella,* (N. Y. Regents Rep. 1859,) and many other forms.

The formation is extensively intersected by divisional planes; and even in those portions not belonging to the brecciated mass at Mackinac, is apt to be considerably broken up. The open character of the rock permits the escape of numerous copious springs of fresh water, and occasionally gives rise to the sudden disappearance of streams and lakes. Various accounts are current, in Monroe county, of subterranean communications from lake to lake, and even between Lake Erie and the western part of the county. I heard it repeatedly stated, that at certain seasons of the year, Ottawa Lake passes off by some subterranean outlet, causing the death of all the fish which remain, but that, on the refilling of the lake, the water is always accompanied by a fresh stock of fish. In Mr. James Cummins' quarry, about five miles, in a right line, north-east of Ottawa Lake, the rock is described as cavernous and full of sink holes; and what is remarkable, is the fact that this quarry is always filled with water when the lake is high, and empty when it is low; and whenever the quarry is full, it contains bass and dog-fish of the common species of that region.

The curious; suture-like structure so often referred to by other geologists, is frequently met with in Michigan—two consecutive layers of rock being studded, on their contact surfaces with tooth-like or prism-like processes which fit into corresponding pits on the opposite surface. A thin film of black bituminous matter generally prevents a perfect contact of the contiguous surfaces. Sometimes these processes are so little developed, that the line of contact is merely zigzag, or truly suture-like, while in other cases they become elongated prisms. The same structure was long ago noted in the same formation, in Ohio, by Dr. Locke.* In New York it is found in the Niagara Limestone, the Waterlime Group, and some of the higher rocks.† These forms were termed by Prof. Eaton, *Lignilites*, from their resemblance to woody fibre. In consequence of Mr. Vanuxem's suggestion,‡ that this structure might be owing to sulphate of magnesia, Dr. Beck subjected to analysis a specimen from the Niagara Limestone, and detected about 21 per cent. of carbonate of magnesia.§

The limestones of this group are generally somewhat bituminous, giving a brownish color and a fetid odor to the rock. The bitumen at Christiancy's quarry in Monroe county, is so abundant as to exude in the form of an oil, and float upon the surface of the water. The bituminous exudation is very marked in the Helderberg limestones of Northern Illinois. Black, bituminous, shaly partings frequently occur between the strata in the upper part of the group.

Considerable hornstone appears in the formation at Raisinville, in Monroe county, and also at Little Traverse Bay. Curious cherty concretions are very common. These sometimes take the form of a perfect sphere, or ellipsoid of revolution, or a gourd, and generally reveal at the centre, traces of some organic substance. These characters are supposed to appertain to the "corniferous" or upper portion of the group. At Brest, Stony Point, Pt. aux Peaux, and some other localities, the broken strata abound in *Strontianite*, *Dog Tooth Spar* and *Rhomb Spar*. At Brest, *Amethyst* is found in limited quantity. Some of the cherty nodules or pebbles at Mackinac, pass to the character of *chalcedony* and well marked *agate*.

The economical importance of this group of rocks is very great. They are everywhere useful for quicklime, and when

*Report of Geological Survey, Ohio, 1838, p. 230.
†Hall, Geological Report, IV District, N. Y., p. 95.
‡Report, III District, for 1838, p. 271.
§Beck, Mineralogy of N. Y., p. 69.

not too remote from settlements, are everywhere burned for this purpose. Monroe county has long been celebrated for the abundance and good quality of its lime. At Christiancy's quarry, about 10,000 bushels are annually produced. The Plumb Creek quarries, below Monroe, furnish a much larger quantity. At Raisinville, where the outcrop of the limestone covers about 200 acres, are 13 kilns, with a capacity of 8,540 bushels. Supposing that these kilns burn, on an average, once in three weeks, the total amount of lime produced is 145,180 bushels per year. Lime is manufactured at numerous other points in the county. According to statistics on hand, the average cost of the lime at the kilns is about 5 cents per bushel. It is sold for 12½ cents at the kilns. Supposing the Plumb Creek quarries to produce 100,000 bushels annually, and all other kilns in the county 50,000 bushels, we have an aggregate of 295,000 bushels, which at 12½ cents a bushel amounts to the considerable sum of $36,875. This lime is generally purchased by the farmers, who carry it in wagons to the surrounding country, for a distance of 30 miles. It is generally sold by them for 25 cents a bushel, making a profit to them of $36,875 which is likewise retained in the county. The aggregate annual addition to the wealth of the county, therefore, from the manufacture of lime alone, is $73,750.*

For architectural purposes, some portions of the Helderberg limestones seem to be extremely well adapted. The sills, caps and water table manufactured at Christiancy's quarry, have a reputation of many years standing. They may be seen in the court house in Monroe, in the new hotel, in all the new block of stores on Washington street south of the city hall, in Wing and Johnson's banking office, and three stores in Monroe street, There is no stone which stands the weather better. They seem even to improve under the influence of exposure. The distance of the quarry from the railroad has, however, prevented these stones from coming into general use. During 1859, twelve

*The scope of this report does not permit further details, though the materials are on hand. They will be introduced into the final report, together with practical suggestions in reference to selecting and burning the stone, and improving the quality of the lime.

hundred feet of caps, window sills and water table were worked out, and about two hundred feet of door sills. About one hundred and twenty cords of rough stone are annually sold for building purposes.

At Crawford's quarry, on the shore of Lake Huron, about eighteen miles beyond Presque Isle, this limestone presents characters which create the hope of very interesting developments. The rock here is compact, fine-grained and handsomely clouded by the unequal distribution of the bituminous matter, so that polished surfaces of the general mass present quite an elegant appearance. The large dome-shaped coral, however, spoken of as occurring at Thunder Bay Island and Little Traverse Bay, produces in the stone at this quarry a very beautiful effect. The undulating concentric laminae, when cut by right planes, and the surfaces polished, exhibit a beautiful agate-like structure, the effect of which is greatly heightened by the coralline disposition of the calcareous matter, and the varied distribution of the bituminous color. Should it be proved that this sort of rock can be procured in samples sufficiently large, the Lake Huron marble will take its place by the side of the most highly esteemed varieties.

The agricultural capabilities of the district underlain by this group of rocks is very great. The whole of the elevated limestone region north of the line joining Thunder and Little Traverse Bays, is capable of supporting a dense population. The contrast noticed in passing from the arenaceous soils of the Marshall and Napoleon Groups, to the calcareous soils of the Helderberg Group, is very striking. The islands of Bois Blanc and Mackinac, but especially the former, are covered with a growth of timber, which, except the addition of a few scattered *Coniferæ*, is a perfect reproduction of the forests of Monroe county, and Northern Ohio. The same might have been said of the plateau upon the Niagara limestone, extending west from Centralia, on Drummond's Island. I saw here the beech, black birch, sugar maple, and other trees growing to an enormous size. One birch measured 10 feet in circumference. Mr. Francis showed me here excellent crops of Indian corn, potatoes and oats.

9.—*Hamilton Group.*

On the east side of Thunder Bay Island, the rocks of the Helderberg Group are seen overlain by a black bituminous limestone, abounding in *Atrypa reticularis,* and numerous other Brachiopods allied to the types of this group. The locality furnishes, also, two or three species of trilobites, a *Favosites,* a large coral allied to *Acervularia* and some fish remains. The rock breaks in every direction, and abounds in partings of dark shaly matter.

The same beds are again seen at Carter's quarry, two or three miles above the mouth of Thunder Bay river, and here it contains the same fossils. It is seen again on the south shore of Little Traverse Bay, replete with Brachiopods and Bryozoa, and is here eighteen feet thick. It is overlain by two feet of dark chocolate colored, compact, argillaceous limestone, much shattered, and abounding in Cyathophylloids and other corals, which, in turn, is surmounted by 14 feet of a limestone varying from calcareous and crystalline to argillaceous, in beds from 2 to 24 inches thick. The whole series is completed by 6 inches of black shale.

The exact order of superposition of all the rocks constituting the Hamilton Group, has nowhere been *observed.* The bluff at Partridge Point, in Thunder Bay, is believed to come in next above the bituminous limestone of the localities just cited. The rock here is at bottom, a bluish, highly argillaceous limestone, with shaly interlaminations, the whole wonderfully stocked with the remains of Bryozoa and not a few encrinital stems. No calices of Encrinites, however, could be found, except two *Pentremites* picked up along the beach, and one *Cyathocrinoid* found in place. Above these beds, which are but five feet thick, occurs a mass of blue shale, six feet thick, calcareous in places, and irregularly interstratified with blue, argillaceous limestone. It contains Bryozoa, Cyathophyllidae and Trilobites.

Still higher, is a massive limestone, below, filled with Bryozoa, Encrinites and Brachiopods, above, little fossiliferous, the whole, with interlaminations of clay.

At the upper rapids of Thunder Bay river, still a different but entirely detached section was observed, and it is, as yet, impossible to collocate it with the others. The same must be said of the isolated exposure at the lower rapids. At the upper rapids (N. E. ¼ of S. W. ¼, sec. 7, T. 31 N., 8 E.,) on the south side of the river, limestone is seen in a bluff 15 feet high, dipping E. S. E., about 5°. The whole section exposed is 25 feet, made up as follows, from above:*

8. Limestone, bluish, flaggy,..........................8 ft.
7. Limestone, dark gray, highly crystalline, thick bedded, with *Favosites*,....................................9 ft.
6. Limestone, dark bluish, very fine grained, hard, compact and heavy, with a few reddish streaks and spots, and some encrinital stems and shells, and a few crystals of spar interspersed, with occasional seams of the same in the form of dog tooth spar. Would make an excellent building stone, and probably would receive a fine polish, ...6 ft.
5. Limestone, gray, crystalline, thick bedded, seen in bottom of river. This rock resembles fragments seen at the highest level about the lower rapids,...........2 ft.
4. An interval of no exposure. Half-a-mile higher up the stream, the section is continued, as follows:
3. Limestone, dark, bluish-gray, fine grained, compact, in layers 2–4 inches thick. Resembles the rock at the lower rapids.
2. Clay, indurated, regularly stratified, rather dark,.....3½ ft.
1. Calcareous shale, with fossils, forming the bed of the river.

The dip at this place is abnormal and evidently local. The true geological position of the rocks must be determined by future investigation.

The rocks of the Hamilton Group are traced from the south shore of Little Traverse Bay to near the outlet of Grand Traverse Bay. At some of the exposures *Spirifer mucronatus* is recognized in great abundance, though by far the most abundant Brachiopod is *Atrypa reticularis*.

*In all the sections given in this Report, the numbering proceeds from below.

The Hamilton Group seems to play a very important part in the geology of the northern portion of the peninsula, but in the southern part of the State it has not yet been satisfactorily identified From Thunder Bay it passes under the bed of Lake Huron, and reappears upon the Canada shore, between Benson's Creek and Cape Ipperwash or Kettle Point. From here, as nearly as can be ascertained from the reports of the Canadian survey, it passes southward in a belt about ten miles wide to the south-eastern part of the county of Lambton, where it is met by another outcropping belt, extending east from the shores of Lake St. Clair. The united belts fill a trough in the Helderberg limestone, which extends east to the shore of Lake Erie between Point aux Pirs and Long Point, whence it crosses the lake, and reappears in Ohio.

The branch which comes in from the direction of Lake St. Clair, ought to be recognized in the southern part of our peninsula, but though we have here a great thickness of argillaceous strata, they are supposed to belong rather to the group above than to this one. It seems, at any rate, pretty obvious that the eminently fossiliferous limestones of Thunder and Little Traverse Bays, do not reach the latitude of Detroit, a fact which accords with the great attenuation of the Helderberg limestones, in the same direction.

In an economical point of view, the rocks of this group have not been shown to possess great interest. It would certainly be well, however, to test the hydraulic properties of some of the argillaceous limestones of Thunder Bay.

10.—*Huron Group.*

At Sulphur Island, in Thunder Bay, not more than a mile east south-east from Partridge Pt., is found a black bituminous slate, which is believed to overlie the fossiliferous cliffs at the latter place. No undisturbed strata are seen on the Island, which consists of a mass of fragments rising a few feet above the water. These slates or shales burn with considerable freedom, and it is stated that a combustion started from camp fires has, in several instances, continued spontaneously for many months, in one case 16 months. The cinders resulting from these fires are still very conspicuous. These shales furnish no fossils, except a few vegetable impressions resembling a *Calamites*, and some very indistinct impressions of shells. Pyritous nodules and septaria are quite common. Capt. Malden, of Thunder Bay Island, gave me a specimen of the latter, in the shape of a very oblate ellipsoid, 14 inches in its greater diameter and 3 in the lesser.

At Squaw Pt., on the main land south of the island, near the residence of the old Indian Chief, Zwanno Quaddo, the black slates are found in place, in a cliff 10 feet high. The exposed surfaces are very much discolored by oxide of iron.

On the opposite side of the State the black shales are seen at the south-east extremity of Mucqua Lake, in Emmet county; on the north side of Pine Lake, (sec. 3, T. 33 N., 7 W.); near the outlet of Grand Traverse Bay, (sec. 3, T. 32 N., 9 W.), and a few miles south of there, and again near the head of Carp Lake, in Leelanaw county. The greatest observed thickness in this part of the State is 20 feet.

On the east shore of Grand Traverse Bay, nearly opposite the north end of Torch Light Lake, is a bed of green shale occupying a position above the black shale. It is rather a soft, semi-indurated clay, traversed by bands of lighter color, apparently calcareous.

No rocks have anywhere been seen reposing upon the black or green shales.

From Sulphur Island, in Thunder Bay, the black shales pass under the bed of Lake Huron toward the south-east, and emerge at Cape Ipperwash, on the Canadian shore. From here they are traced to the township of Mosa, in Middlesex county, and, from their occurrence at Enniskillen and other localities in the vicinity, they may be regarded as occupying the triangle embraced between the two belts of Hamilton rocks, before referred to, and the National boundary line. This triangle would be the thinning out corner of the great basin which forms the Lower Peninsula of Michigan.

These shales, at Enniskillen, Bear Creek and neighboring localities in Canada, become the source of large quantities of petroleum; and there is little doubt that the mineral oil of Ohio is derived from the same formation.* These shales, and the great mass of less bituminous shales lying above them, contain a vast amount of vegetable or animal matter, the source of the rock oils. This oil is eliminated by a slow spontaneous distillation, and rises up and saturates the overlying porous sandstone rocks, in which, in Ohio and Pennsylvania, it is found by boring.

Does the rock oil exist in Michigan? The oil bearing rocks of Enniskillen, are but an elbow of a formation which belongs properly to the Michigan side of the boundary line. The oil producing shales unquestionably dip under our State, and are not far from the surface throughout St. Clair, Oakland, Macomb, Sanilac and Huron counties But are they overlain by a porous sandstone capable of becoming the repository of the products of the spontaneous distillation of the oil, or are they overlain by argillaceous strata which would prove completely impervious to the ascent of volatile matters? In the present state of our knowledge this question cannot be satisfactorily answered, but the indications are not altogether favorable. Nevertheless it is well known that at several points in St. Clair county evidences of bituminous exudations exist, and streams of inflammable gas have escaped from the earth; moreover, an overlying sandstone does not seem to be everywhere an essential condition to the accumulation of oil. In the present state of the case there seems to be sufficient encouragement to embark in explorations on a cautious scale.

The strike of the black bituminous shales beneath the bed of the lake, from Thunder Bay to Kettle Point, must pass several miles to the east of Point aux Barques. It follows, therefore, that the shales and flagstones occurring along the shores of Huron county and dipping toward the south-west, must be many

* See an interesting paper on the "Rock Oils of Ohio," by Dr. J. S. Newberry, extracted from the Ohio Agricultural Report for 1859.

feet higher than the shales of Thunder Bay and Kettle Point. The Huron county shales and flagstones, however, are the next rocks observed in ascending order. Not less than 180 feet of them, are seen in Huron county, and the total thickness must be much greater. They were penetrated 59 feet in Butterworth's salt well at Grand Rapids, 130 feet in the State salt well, and 214 feet in Lyon's well.

The greater part of this member of the group consists of shales, which are laminated, fissile, dark blue or blackish, bituminous and pyritiferous. Their exposed surfaces generally become covered with rust, and when protected from the weather, with an astringent eflorescence resembling sulphate of iron. Throughout the whole thickness, we find occasional bands of hard limestone and bluish, fine-grained, somewhat argillaceous sandstone, which at many points has been manufactured into whet-stones, and might be used for flagging. The more shaly portion is surmounted by a more important mass of the sandstone fifteen feet in thickness, from which the celebrated Huron grindstones are manufactured. The rock here is bluish-gray, fine-grained, perfectly homogeneous, with sharp grit and a limited amount of argillaceous matter. Between the layers are found some serpentine grooves and casts like worm tracks. One of these was traceable twenty-eight inches and was three-eighths of an inch in width. In one fragment the pectoral fin of a fish is preserved. Numerous obscure traces of terrestrial vegetation are found between the strata, and in one place the workmen opened a cavity from which they took out a bushel of good bituminous coal—a discovery which was immedately followed by a fever!

The junction between the gritstones and underlying shales, is finely seen at the old quarry, about one mile east of the principal one, the upper fourteen feet being sandstone, and the the lower six, shale. It is again seen at the mouth of Willow Creek, where, near the saw mill, the shale rises six feet, and is overlain by the gritstone. The latter is struck in all the wells of the neighborhood, and forms a high ridge to the east of the village. Following up the creek for two and a half miles the land is found to rise rapidly, and the banks of the creek are in some places sixty to eighty feet high. The elevation here rises up into the group next above.

At the light house, one mile east of Willow Creek, the following section is seen:

9 Shale, with interlaminations of sandstone........... 12 ft.
8. Sandstone, bluish, fine,............................ 2 ft.
7. Arenaceous shale,................................. 2 ft.
6. Sandstone, bluish, hard, concretionary,.............. 2 ft.
5. Shale, very persistent,............................ 3 in.
4. Sandstone, calcareous, hard, highly fossiliferous; contains *Retzia*, *Merista*, *Gomphoceras* (?) *Clymenia*, *Rhynchonella*, a *Spirifer* resembling *S. mucronatus* and *S. medialis*, but distinct from both, and a large Leptænoid shell, $2\frac{3}{4}$ inches across the hinge line,.......... 2 ft.
3. Shale,.. 2 ft.
2. Sandstone, hard, pyritiferous, very persistent,........ $1\frac{1}{2}$ in.
1. Shale,.. 12 ft.

The hard, projecting, pyritous layer, (2) affords an excellent opportunity for measuring the dip of the formation, which was found to be one and a half degrees toward the south-west.

A short distance west of the light house occurs the most extensive dislocation seen south of Mackinac. In the neighborhood of the disturbance, on each side, the strata exhibit short undulations, which finally become an actual break, and downthrow of five or six feet. Indications of a sliding movement are seen in the vicinity, and the whole effect is such as might be produced by a lateral pressure from the west.

The gritstones of Lake Huron are destined to play an important part in the economical geology of the Lower Peninsula. The principal quarry owned by Johnson, Pier and Wallace (sec. 30, T. 19 N., 14 E.) is now worked over an area of little more than four by twelve rods. Two hundred tons of grindstones were taken out during 1859, and I was informed by the foreman that he expected to manufacture five hundred tons during 1860. Several stones have been finished, weighing a ton each, and one which weighed three tons. These facts show the soundness and homogeneous character of the formation.

For flagging, and for window caps, sills and water-tables, this stone is equally adapted. When wrought, it has much the appearance of the Waverly sandstone. Its color is decidedly preferable to that of the freestone, so extensively introduced from Cleveland, Ohio. It contains less ferruginous matter, and is less likely to stain.

The outcrop of the shales of this group is seen in the southern part of the State, near Adrian, in Lenawee county; near Union City, and again near Coldwater, in Branch county; at Athens, Leroy, and Newton, in Calhoun county; at Mendon and Leonidas, in St. Joseph county; and at Bangor, in Van Buren county. There is little doubt that the low argillaceous belt of country between Adrian and the region west of Detroit, marks the continuation of the outcrop of the same rocks. An Artesian well bored at Detroit 1829–30, showed the existence of 118 feet of plastic clay overlain by 10 feet of soil and sub-soil, and underlain by 2 feet of sand and gravel resting on solid limestone. It has already been stated that the shales were penetrated in three of the salt wells at Grand Rapids. At the well of Hon. Lucius Lyon, the boring extended 214 feet into these strata, without reaching the bottom. This boring passed a 2 feet band of sandrock 18 feet from the top of the shales, and a 1 foot band 50 feet from the top—the arenaceous element being thus shown to be much less abundant than in Huron county. The shales were penetrated 130 feet in the State salt well, 3 miles west of Grand Rapids, and 59 feet in Butterworth's well.

In Branch county the shales, or more properly clays, are freighted with a considerable abundance of kidney iron ore, which was formerly used, to a limited extent, in the furnace at Union City, but found too highly charged with sulphur to answer well.

At two localities—Leroy, in Calhoun county, and Mendon, in St. Joseph county—these argillaceous beds present the character of a black bituminous shale. In Mr. Canwright's well, near Coldwater, the upper part is also bituminous, but soon passes into a plastic dark blue clay, which he has worked very extensively in the vicinity, in the manufacture of bricks. For this use, the kidney iron clays are generally well adapted

No fossils have been detected in this group in the southern part of the State, except a *Tellina*, a *Solen* undistinguishable from one in the Marshall Group, a *Chonetes* and a *Grammysia*.

The bituminous character of most of the shales of this group, and especially of that portion known as the "Black Bituminous Shales," has given rise to numerous misapprehensions in regard to their geological relations, and has been the occasion of the practice of a great amount of geological quackery. The popular opinion is, that coal *must* exist somewhere in the vicinity of the black shales. The opportunity has been very many times presented for discouraging explorations contemplated or undertaken, under the influence of this illusion. Large tracts of land have been secretly taken up, with the view of securing eligibly situated coal mines. The reports so rife among the Indians and their missionaries, of the occurrence of coal in the neighborhood of Grand Traverse Bay, are undoubtedly traceable to the same illusory shale. There is not the remotest probability of the occurrence of coal within a hundred miles of Grand Traverse Bay. This statement is made in full recollection of the allegation of a learned judge, that he had seen *anthracite coal* that was said to have been collected in that region. One of the localities, of Indian notoriety, is at the southern extremity of Mucqua Lake, south of Little Traverse Bay. The Indians report that they have often resorted there for fuel, and that they have burned the *coal* in their camp fires—a statement perfectly credible if we substitute *shale* for *coal*.

Similar misguided expenditures have been made in the same rocks in Canada, New York, Ohio and other States.

The geological positon and equivalents of the Huron Group of rocks, cannot yet be regarded as satisfactorily settled, and for this reason they have received a provisional, local name.

The black bituminous slate of Michigan has generally been regarded as equivalent to the "black slate" of Ohio and Indiana, which is reputed to occupy the horizon of the Marcellus shale or perhaps the Genesee slate of New York. The Marcellus shale, however, lies below those New York rocks whose equivalents are found at Partridge Point, while our black slate lies above, more nearly in the position of the Genesee slate, or some of the shales of the Hamilton Group. The lithographical resemblances, as inferred from the New York Reports, seem to give color to this identification.

With reference to the setttlement of this and similar geological questions, I paid a visit to several localities in the vicinity of Cleveland, where observations have been made by Dr. Newberry, Prof. Hall and others. Dr. Newberry accompanied me to several points and rendered me every possible assistance. About 3 miles east-south-east from Cleveland is an outcrop of sandstone dipping south-east. This is at top, coarse, glistening and somewhat mottled. Below, it becomes light colored, then dirty reddish gray, and then highly ferruginous, with ironstone partings. On the whole it closely resembles the sandstone of the upper part of the Marshall Group. It is said by Dr. Newberry to be 150 to 200 feet below the conglomerate. At Mecca, in Trumbull county, it is completely saturated with oil.

At a lower level I observed chocolate colored or reddish shales with interlaminations of light blue, argillo-calcareous slate. From the equivalent of these shales on the west side of the Cuyahoga river, is manufactured the mineral paint of Ohio

Still lower, were noticed beds of concretionary shale, or flagstones, underlain by fissile shale. The under surfaces of the former are marked by the appearance of flowing mud, a phenomenon described as occurring in the Portage sandstones of New York.

At a still lower level occurs a large stone quarry, showing a section through a series of bluish, fine-grained sandstones with shaly partings from half an inch to a foot thick. These beds very closely resemble the Huron county gritstones, and are regarded by Dr. Newberry as the base of the Portage Group in Ohio.

Further down the ravine are seen twenty or thirty feet of dark fissile shales, covered with iron rust and an astringent efflorescence, and in every respect resembling the shales which underlie the gritstones of Lake Huron. Unfortunately there is no possibility of founding an equivalency on palæontological evidence. Aside from this I am constrained to regard the flagstones and shales of Cleveland as on the horizon of the gritstones and shales of Lake Huron. But the Cleveland shales are regarded by Dr. Newberry as "Hamilton shales," perhaps, however, using the term Hamilton in the extended sense, so as to include all the New York strata from the Marcellus to the Portage. If the overlying shales and flagstones of Lake Huron, and the underlying argillaceous limestones of Partridge Pt. fall into the Hamilton Group, the intermediate black bituminous shales occupy the same position. So I had been inclined to regard them So I subsequently learned the black shales of Enniskillen were at first regarded by Mr. Billings, though he afterwards placed them in the Portage Group on the judgment of Prof. Hall. This palæontologist, whose authority is not to be questioned where palæontological evidence is within reach, thinks he likewise recognizes in the vegetable impressions of the black shales of Michigan, and in their general physical characters, satisfactory affinities with some of the shales of the Portage Group. In this state of the case we shall be constrained for the present to regard the Huron Group of Michigan, extending from the conglomerate above the gritstones of Huron county, to the top of the argillaceous limestones of Partridge Pt., as probably representing the rocks of the Portage Group of New York.

From the description which has been given of the Huron Group in its northern and southern outcrops, it appears that the group is composed of coarser materials toward the north, and probably attains in that direction, much the thickest development, while, in the State of New York, the source of the materials seems to have been from the east.

11.—*Marshall Group.*

In Huron county, we find the gritstones separated from the higher sandstones by a conglomerate about two feet in thickness, in which occur some of the fossils of the overlying group, especially a *Rhynchonella* of undescribed species, which, in some localities, forms entire masses of rock. From the grindstone quarries to Point au Chapeau, the coast is occupied by sandstones which, at the various "Points" rise in bluffs from eight to twenty feet high, and farther back from the shore attain, in some instances, considerable elevations. The distinction between the Marshall and Napoleon Groups is not clearly traced along this coast. At Hard Wood Point, three-fourths of a mile west of Pt. au Pain Sucre, (called also Flat Rock Point,) are seen, proceeding from the west, the first undoubted fossils of the Marshall Group. The rock here, which rises but a few feet above the surface, is a fine grained, bluish sandstone, with minute glistening scales of white mica. It embraces a *Nucula* characteristic of the Marshall sandstone, a *Solen*, a *Clymenia* and a *Goniatites*. The *Clymenia* occurs in a purplish, fine grained sandstone of exceeding hardness, equaling, in this respect, the Medina Sandstone. In a specimen of the rock found here, containing carbonaceous specks, were seen small geodes lined with rusty crystals of calcareous spar, and containing small imbedded crystals of *native copper.*

Between this locality and Flat Rock Point, the section near the shore reveals several feet of purplish, greenish and yellowish strata, successively lower in the series, in some of which I recognized a minute *Cypris*-like shell similar to one seen at numerous points in the southern part of the State. At Flat Rock Point, still lower rocks rise ten feet above the water, characterized by oblique laminæ of great extent and uniformity, dipping 45° toward N. 38° E. The whole rock here is a purely quartzose, friable sandstone, with many disseminated small pebbles.

From this place to the immediate vicinity of Port Austin, rocks lower and lower in the series rise to the surface, frequently attaining an elevation of 12 feet or more. The first of the series is a bluish gray sandrock, 12 feet thick, followed by a whitish and grayish, sometimes yellowish, fine grained sandstone, very pure and massive, occurring in beds 10 to 12 feet thick, without pebbles or seams, and moderately coherent. At the point one mile west of Port Austin, it is broken into immense angular fragments forty feet and less, in diameter, which lie about like the work of Titanean quarrymen. Immense chasms produced by fissures through the rock, extend inland several rods, and in some cases return again to the water, thus detaching areas a quarter of an acre in extent, and even more. Upon these rocks are growing the Red Cedar, Hemlock, *Pinus resinosa*, Arbor Vitæ or White Cedar, White Birch, Wintergreen and extensive beds of the delicate little *Linnœa borealis.*

At Pt. aux Barques, is seen a sandrock dipping south-west $1\frac{1}{2}^{\circ}$ and consequently passing beneath the last. The outcrop exposes 12 feet. The lowest beds here are red-striped sandstone, similar to some parts of the Marshall Group, in Calhoun and Hillsdale counties. Farther along, on the most projecting part of the point, the striped sandstone rises four feet above the water, and in the immediate vicinity, the cliffs attain the heighth of 17 feet. This is by the Trigonometrical Station of the Lake Survey. The overhanging cliffs here, seen from a distance, bear a rude resemblance to the prow of a vessel projecting over the water, and suggested to the early navigators the name which is still borne by the point, and to some extent attaches itself to the whole region for several miles east and west.

At the fishing station and residence of J. G. Stockman, half a mile east of Pt. aux Barques, I saw a fine specimen of highly ferruginous sandstone, completely filled with fossils, among which occurs a *Rhynchonella* (n. sp.) and the *Bellerophon*, so abundant in the Marshall sandstone, which I have named *B. galericulatus*.

At the first small point east of Burnt Cabin Pt., a greenish blue sandstone is seen rising to the surface and forming a bluff 8 feet high. This rock contains the *Clymenia* of the grindstone quarries, a mile further east, and with care may be traced to that point where it is found overlain by a conglomerate 2 feet thick, apparently forming the base of the group.

Such is a general description of the sandstones of the coast of Huron county, from the highest beds containing *Nucula* to the conglomerate above the gritstones, both included. It has not been deemed proper to occupy space with the details of stratification at the several points at the present time.

The rocks of this group, as well as those of the Napoleon and Huron Groups, should make their appearance again on the opposite side of Saginaw Bay, between Thunder Bay and Ottawa Pt. This whole coast is, however, destitute of a single outcrop. Nevertheless, the great accumulation of sand along the beach, and the well known arenaceous character of the country further west, affords a sufficiently strong presumption that the limits stated cover the place of outcrop of these groups.

In Sanilac county, near the head waters of the Cass river, sandrock is exposed to a considerable extent, which undoubtedly belongs to the Huron county series. On the S. E. ¼ sec. 7, T. 13 N., 12 E., are found numerous fragments of a coarse, gray, micaceous sandrock, sometimes inclining to greenish, and sometimes mottled or striped with red. Many of these fragments contain white quartzose pebbles, and the whole aspect of the rock recalls that seen at Pt. au Pain Sucre. From this point actual outcrops are frequent as far down the stream as the line of Tuscola county, and even to S. W. ¼ sec. 1, T. 13 N., 11 E., where it rises 5½ feet above the water. The general character of the rock is shown by the following section on sec. 7, T. 13 N., 12 E.:

10. Sandstone, coarse, thin bedded and quite soft, (545 A—F.)
9. Flaggy sandstone, (545 G.)
8. Thin shaly sandstone, passing down to a sandy shale, containing much carbonaceous matter, and with occasional partings of a substance composed of sand, clay and carbonaceous matter finely comminuted, (545, H—L.)
7. Sandstone, shaly and flaggy, (545, N.)
6. Sandstone, flaggy, striped with red, (545 O, and 544.) Interval of 40 rods, up stream.
5. Sandstone in thin layers, (543, A—B,)..............20 in.
4. Sandstone, thick bedded, mottled with red above, striped below, (543, C—D,).................... 4 ft
3. Sandstone, with quartz pebbles, (543, E—F)
2. Sandstone, thin bedded, (543, G.)
1. Sandstone, coarse, soft, very ferruginous, (543, H.) Interval of 30 rods to collection of fragments before referred to.

From this neighborhood to Jackson county, no outcrops of rock are known; but the arenaceous character of the drift materials through Lapeer and Oakland counties and portions of St. Clair and Macomb, renders it not improbable that the arenaceous strata of the Marshall and Napoleon groups would be found underlying that region.

In the southern part of the State, the Marshall Group is better characterized and more fully distinguished from the Napoleon Group above. Throughout all the northern part of Hillsdale county, we find a series of highly ferruginous sandstones, generally very fossiliferous, and easily recognized. The ferruginous matter is often collected into bands of iron-stone, from one-fourth of an inch to four inches thick, sometimes horizontal, sometimes oblique and sometimes concretionary in their arrangement. From a brick red sandstone the rock varies to pale red, yellowish and buff; and lower down, becomes yellowish-green, reddish-green, bluish-green and bluish. At the lowest points, as in Noe's well at Jonesville, it becomes a bluish, micaceous, thin-bedded, shaly sandstone, and thus passes into the shales of the Huron Group below.

Good exposures of the formation may be seen in the quarries

at Jonesville and Hillsdale, and at many points in the townships of Moscow and Scipio. In Jackson county the formation extends up into Liberty and Hanover, and has been pierced nearly through at the depth of 105 feet in the well of S. Jacobs, Jr., in the township of Pulaski. The most characteristic outcrops are found in Calhoun county; and from that at Marshall, the group has received its provisional name. At this place the stratification is as follows:

4. Sandstone, rather thick-bedded, reddish, ... 10 ft.
3. Sandstone, dark-reddish, rather hard, very fossiliferous, 5 ft.
2. Sandstone, reddish-green, homogeneous, thick-bedded,. 10 ft.
1. Sandstone, light, greenish-gray, thick-bedded.

Several characteristic outcrops occur in the township of Marengo, Calhoun county. At Battle Creek the lower beds of the group are seen in places, highly calcareous and very hard, but filled with characteristic fossils. The formation has not yet been seen in place in Kalamazoo and Allegan counties, but numerous fragments of a purple sandstone are strewn over the surface, identical in general aspect with some layers of the group at Pt. au Chapeau, on Lake Huron. In Ottawa county the group presents well marked exposures at several points on sec. 21, T. 5 N., 15 W.—township of Holland. I am also informed by Henry D. Post, Esq., of Holland, that an outcrop occurs in T. 5 N., 16 West., near the shore of Lake Michigan. At these points it embraces, as usual, the characteristic fossils. One mile east of Eastmanville, on the wagon road from Grand Haven to Grand Rapids, a cut in the valley of Deer Creek exposes the laminated areno-argillaceous strata belonging to the lower part of the group; and where the same road crosses Sand Creek, about four miles east of Lamont, numerous fragments and other indications of the neighborhood of an outcrop may be seen. In some of the fragments, which are highly ferruginous, I found the best preserved fossils that I have seen in the State, including *Nucula*, *Orthis*, *Chonetes* and *Orthoceras*.

Further north than this, the group has not been traced; and even to this point, the boundaries are poorly defined, in consequence of the drift materials strewn over the surface, and the perishable nature of the rock. From what has been said, it appears that this group touches Lake Michigan, and that the Huron and Hamilton Groups (if both exist) must pass entirely beneath the lake, re-appearing probably in Mason, Oceana and Manistee counties, while the Marshall Group proceeds in the direction of Newaygo and Lake counties.

Details of stratification and fossils at the various outcrops cannot, of course, be appropriately given at the present time, nor even an enumeration of all the outcrops.

The palæontology of the Marshall Group possesses considerable interest, both in consequence of the number of individuals and species found fossil, and the distinctness of the fauna from that of other regions in the same geological horizon. Considerable attention has been bestowed upon the collections from this group, but not a single satisfactory identification has yet been made. The most abundant and characteristic fossils at the various localities belong to the genera *Nucula*, (5 species,) *Solen*, (2 species,) *Bellerophon*, (3 species,) *Orthoceras* (5 species), *Myalina* and *Clymenia* (5 species). Besides these, I have referred to *Cyrtoceras*, 4 species, *Cryptoceras*, 2 species, *Trocholites*, 1 species, *Goniatites*, 5 species, *Pleurotomaria*, 1 species, *Tellina*, 1 species, *Cardium*, 2 species, *Lucina*, 1 species, *Chonetes*, *Orthis* and other Brachiopods, one or more species each.

There are, moreover, numerous species which have not yet been particularly examined, among which are a few fish remains and land plants. As I intend communicating to the public at an early day, further particulars regarding this assemblage of fossils, I refrain from extended remarks at the present time. The delay experienced, however, in printing this report, enables me to append a few observations relative to the Clymeniæ. According to all authorities, the two genera Clymenia and Goniatites are widely distinguished by the position of the siphon, being interior in the former and exterior in the latter. It is true that all my specimens of Cephalopods from the Marshall Group are rather imperfect; but I have had the opportunity to examine a large number of transverse sections of the so-called Clymenia, and in every case I find indications of a siphon closely internal, while in an equal number of cases, the best possible observations upon the dorsal surface have failed entirely to disclose a siphon in this position. At the same time, it must be admitted that some of Sandberger's figures of "Goniatites" present a close resemblance to some of my Clymeniæ—for example, Figs. 13, 14a Taf. III., and 11c Taf. VIII. Even the sectional view, 11a, Taf. VIII., presents much the appearance of some of my specimens; but while the specimen here figured may have a dorsal siphon, my own specimens have not.

Further, many of the Goniatites (now so-called), figured by DeKouinck, afford to my eye no indications of an external siphon. I have specimens from Rockford, Ind, generally reputed identical with DeKoninck's *G. rotatorius* and *G. princeps*, (properly *G. Ixion* and *G. Oweni*, *Hall*,) and while I admit that the latter has a distinct dorsal siphon, I confess that the former seems to me to have a distinct ventral one!

Such were my convictions at the time of Prof. Hall's visit to Ann Arbor, near the close of November last. In view of the contradictions, I showed him some of my specimens, and without making a critical examination, he did not dissent from my conclusion as to their generic relations. More recently, however, in a letter accompanying a copy of his "Contributions to Palæontology," for 1858–9, and '60, he says, with reference to specimens in his possession from New York and Indiana: "On reviewing my specimens after my return home, I do not find reason to doubt their Goniatitic character." And with reference to my specimens, he adds, "The appearance of siphuncle on the ventral side, which you pointed out, is, I think, deceptive." In accordance with this view, he has referred to Goniatites all of the closely coiled Cephalopods, characterized in this last number of his "Contributions." If, on careful examination of my specimens, Prof. Hall should pronounce them Goniatites, I should yield to his judgment. But the shells in question seem to my eyes to belong to Clymenia, and I can do nothing but regard them as such until I am convicted or positively contradicted.

I cannot doubt that the palæontological characters and stratigraphical position of the Marshall sandstone place it conclusively above the horizon of the Hamilton Group; and hence I am not surprised that none of the nine species of Goniatites described by Prof. Hall, and referred by him to the Hamilton Group, bear any considerable resemblance to the Michigan fossils under consideration.

From this group were collected, at Battle Creek, the specimens described by R. P. Stevens,* as *Leda dens-mammillata, L. nuculæformis, L. pandoræformis, Nucula Houghtoni* and *Chonetes Michiganensis.* Not one of the Lamellibranchs has been satisfactorily identified by me, amongst the fossils collected at the same locality. The Nuculoid shells have not the pallial sinus nor posterior elongation required by their assignment to the genus *Leda*; nor, supposing them true *Nuculæ*, do I find their specific characters clearly indicated. Moreover, Dr. Stevens' reference of these fossils to "ochreous shales, belonging to the coal measures," because "associated with an *Orthoceras*, a *Nautilus* and *Bellerophon Urei*, which is evidently carboniferous," must undoubtedly be regarded as an oversight. The occurrence of *Clymenia* in these rocks establishes their Devonian age, while the *Bellerophon* supposed to be the one referred to, is quite distinct from *B. Urei* of Fleming, which is a dorsally sulcated shell, while ours presents no trace of such a character. Still further, *B. Urei*, even if occurring here, would not identify these rocks with the "coal measures," since the range of this species is from the Upper Silurian to the Mountain Limestone.

The general aspect of the fauna of the Marshall Group bears some resemblance to that rep esented by the figures of the fossil remains of the Rhenish Provinces of Nassau,† in Ger-

*Silliman's Journal, Vol. XXV, [2] p. 262.

†See Sandberger's Systematische Beschreibung und Abbildung der Versteinerungen des Rheinischen Schichtensystems in Nassau.

many, though we have not so large a proportion of *Goniatites;* while *Trilobites* and *Spiriferidæ* are entirely wanting. Neither is our fauna by any means as rich.

The rocks of this group have been quite extensively employed in the southern part of the State for building purposes, and in moderate sized structures they answer sufficiently well, but for very high structures the stone needs to be selected with care, as some portions are too incoherent for security. For cellar walls and other rough masonry they prove of great utility. At Jonesville and other localities the uniformly colored, homogeneous, greenish strata, in the lower part of the group, have been worked into very handsome caps and sills.

12.—*Napoleon Group.*

In approaching Point au Chapeau of Lake Huron, from the south, the bottom of the lake is seen to be a solid greenish sandstone. At the point is an outlier containing about four square rods. The section exposed here is about 8 feet. The action of the waves has undermined the rock, and excavated it into purgatories through which the water rushes with the hollow sound described as occurring in similar situations on the sea coast. In one of these purgatories the following section was observed:

4. Sandstone, reddish gray, with rusty specks, and many coarse grains of white quartz.
3. Sandstone, very thinly laminated, fine-grained, and of a dirty greenish color.
2. Sandstone, yellowish-red, with conspicuous grains of white quartz, and particles of rusty matter.
1. Sandstone, reddish and otherwise similar to above

All the strata exhibit oblique and curved lamination, the dip of the oblique laminæ being at this place toward the north, at an angle of 45°. Between (2) and (3) are thin layers of bluish micaceous, carbonaceous sandstone of local occurrence.

On the east side of this point, the overhanging cliff has formed a sheltered cave, in which, with some additions from rude art, it is said an old hermit found a tolerable habitation for several years. The ruins of his stove are still visible.

It must be confessed that in lithological characters, these rocks cannot be distinguished from strata of the Marshall Group; and they are assigned to the Napoleon Group simply in consequence of their occurrence at a higher geological level than the highest strata, (those at Pt. au Pain Sucre,) which contain *Nucula* and *Clymenia.*

There is little doubt that some of the sandstones before referred to, as occurring near the forks of the Cass river, should be assigned a position in this group, but it is impossible with our present knowledge, to draw dividing lines.

The next outcrop of these rocks is found at Napoleon, in Jackson county, where they are quarried over an area of 88 acres, and expose a section of about 75 feet. The rock is for the most part of a grayish color, inclining to buff. The beds are generally of sufficient thickness and perfection to answer either for flagging or building. The following is the stratification:

4 Sandstone, buff and bluish-gray, composed of transparent and colored grains of quartz, thick bedded,.........40 ft.
3. Sandstone, yellowish, thick bedded,................. 4 ft.
2. Sandstone, thick bedded, pale greenish,.............20 ft.
1. Sandstone, greenish-buff, composed of minute rounded grains of colored quartz pretty firmly cemented with a very perceptible quantity of white calcareous matter, ...11 ft.

The higher beds are worked on the grounds into excellent window sills and water-tables, which sell for 28 cents per linear foot. I saw some fine floated and moulded stone steps and door-sills, selling for 37½ cts. a square foot. The rough stone costs 25 cts. a perch at the quarries, or 50 cents a perch of 1600 lbs. on board cars. The charges for freight are two cents a hundred to Jackson, four cents to Adrain, ten cents to Monroe. The quarries at this place furnished the cut stone for the new Union School building in Monroe, also for the City Hall at that place, and the Union School House at Tecumseh. A fine store of this stone, with smooth front, has been erected at Hillsdale.

Some beds of this stone are sufficiently clean and sharp to

answer the requisites of a coarse grindstone, and some years ago this manufacture had attained here a considerable degree of importance.

The Napoleon sandstone outcrops at numerous other localities in the south part of Jackson county, and further northwest. Being entirely destitute of fossils, it is not easy to distinguish it from the sandstones above, and the unfossiliferous portions of the sandstones below. The most northern exposure yet examined on the southern slope of the State, is in the right bank of the Grand River about a mile above Grandville, in Kent county.

In all the borings for salt which have passed through the Napoleon sandstone it has been found separated from the Marshall Group by a bed of clay. This, at the State salt well, was 14 feet thick; at Lyon's well, 9; at Butterworth's, 10; at Scribner's, 10; at the Indian Creek well, 15; at Windsor's, 10; at East Saginaw, 64 feet. The thickness of the overlying sandstones is pretty uniformly about one hundred feet.

The Napoleon sandstone bears considerable resemblance to the conglomerate of Ohio, as seen in the gorge of the Cuyahoga, at the falls; but it contains no pebbles, and occupies a position, moreover, below the carboniferous limestone. As a distinct formation, therefore, it has no satisfactory equivalent in surrounding States; and there is no reason, except its negative palæontological characters, for separating it from the Marshall Group. The uniformity in the petrographic character of the sandstones of Huron county, has already been alluded to. Should it hereafter appear that the separating shale which lies between the Marshall and Napoleon Groups of the southern part of the State is wanting in the north, we shall be obliged to regard the one hundred and nine feet of sandstone passed in the deep well at East Saginaw as representing both these groups, diminished to the thickness of one of them; while the shale beneath, penetrated to the depth of 64 feet, must be regarded as the commencement of the argillaceous portion of the Huron Group. Such a thinning of strata toward the north would, however, constitute a reversal of the general law of our strata, and I have consequently been induced for the present to regard the shale reached in the Saginaw deep well as the thickened separating shale lying between the Napoleon and Marshall Groups.

13.—*Michigan Salt Group.*

The Napoleon sandstone, exposed along the right bank of the Grand River a mile or two above Grandville, in Kent county (S. E. ¼ sec. 7, Wyoming), near the residence of Mr. Davidson, is succeeded upwards by a remarkable series of saliferous shales and intercalated beds of gypsum and magnesian limestone, attaining a maximum observed thickness of 184 feet. The lower portion of this formation outcrops in an extensive salt marsh, on sec 3, T. 6 N., 12 West (Wyoming, Kent county). This is the locality of the State salt well, near Grand Rapids. Nearly opposite, on the north side of the river, in a bluff rising 60 or 80 feet above the water, are located extensive gypsum quarries. At the quarry known as McReynolds & Stewart's, I observed the following section:

19.	Loam, variable in thickness.	
18.	Clay, yellowish and plastic,	3 ft.
17.	Shale,	3 ft.
16.	"Plaster rock"—a series of irregularly alternating layers of arenaceous limestone and shale, inclosing many masses of reddish gypsum,	5 ft.
15.	Limestone, argillo-arenaceous (called "flint,")	4 in.
14.	Shale, blue, thinly laminated, pretty uniform,	4 ft. 6 in.
13.	"Water limestone,"	8 in.
12.	Shale,	1 ft.
11.	Water limestone,	10 in.
10.	Shale,	3 ft.
9.	"Plaster rock," composed of plaster, with some clay,	2 ft.
8.	Shale,	3 ft.
7.	Water limestone, (which in Hovey's quarry was found to pass into gypsum,)	2 ft.
6.	Shale,	3 ft.
5.	Gypsum,	6 ft.
4.	Shale,	9 in.
3.	Gypsum,	13 ft. 6 in.

2. Shale,	2 ft.	
1. Limestone and gypsum, more than	4 ft.	
Total,	57 ft.	7 in.

The following is the section at Hovey & Co.'s plaster quarry within a few rods of the last:

16. Loam,	6 ft.	
15. Clay,	3 ft.	
14. "Water limestone,"	1 ft.	
13. Shale,	4 ft.	
12. Gypsum,		10 in.
11. Shale,	1 ft.	3 in.
10. Water limestone and clay in thin layers,	2 ft.	
9. Shale,	3 ft.	
8. Gypsum,	1 ft	6 in.
7. Shale,	3 ft.	
6. Water limestone,		10 in.
5. Shale,	4 ft.	
4. Gypsum,	6 ft.	
3. Shale,	1 ft.	3 in.
2. Gypsum,	13 ft.	
1. Gypsum, hard, rather dark colored, through which the excavations have not yet extended.		
Total,	44 ft.	8 in.

In establishing a parallelism between these two sections, it is probable that we must regard Nos. 1 and 2, (Hovey,) as the equivalent of No. 3, (McReynolds;) No. 3 (H.)=4 (McR.) &c.; No. 13, (H.) corresponding to 14 (McR.); 14 (H.) to 16 (McR.); 15 (H.) to 18 (McR.); so that the beds 15 and 17, (McR.) find no equivalents in Hovey's quarry.

The 13 feet bed of gypsum is a pure and solid mass. At top it is reddish, veined with the bluish color of the shale; below, it becomes more bluish as a mass. At the center the fracture and lustre remind one of hornstone, the mineral being translucent, fine grained, compact and homogeneous. From this to the bottom of the bed, is a mottled and clouded gypsum of a coarsely fibrous structure.

The shales of McReynolds & Stewart's quarry are said to

effloresce with common salt in dry weather and furnish a favorite "lick" for cattle.

The roof of McReynolds & Stewart's quarry dips N. W., about one foot in twenty. The dip in Hovey's quarry is very slight, N. 10° E.

By connecting these observations with those made in boring the State salt well on the opposite side of the river, we arrive at an approximation to the whole thickness of the group, thus:

Section measured at McR. & S.'s quarry,..............	58 ft.
From bottom of quarry to alluvial flat by river's edge,..	50 ft.
Allowance for dip of formation,......................	15 ft.
Thickness of alluvium at salt well,..................	40 ft.
Residual thickness of salt strata in well,..............	21 ft.
Total,..	184 ft.

This series of rocks is penetrated in all the borings for salt, at Grand Rapids and that vicinity. It is found passing upwards through a few feet of calcareous sandstone, into the well characterized carboniferous limestone. The thickness of the group in Lyon's salt well, was found to be 171 ft.; in Butterworth's, 157 ft.; in Scribner's, 153 ft.; in the Indian Creek well, (Ball's) 133 ft.; in Windsor's, 184 ft.; and it was penetrated 100 feet in Powers & Martin's well. In Jackson county it is found to be 49 feet thick.

In Kent county, the Michigan Salt Group is undoubtedly the source of the supply of brine, though the strength remains undiminished, as a matter of necessity, while the boring is continued in the underlying Napoleon sandstone, until a stream of fresh water is struck, which, rising up, materially dilutes the brine

On the opposite side of the State, this group outcrops on the shore of Tawas Bay (Ottawa Bay), on the west side of Saginaw Bay. Two miles beyond White Stone Pt., Bay county, T. 20 N., 7 E., is a bluff about 19 feet high, known as "Plaster Bluff," at which the following section was observed:

H. Clayey subsoil.
G. Limestone, thin-bedded, resembling E., ... 4 ft.
F. Sand, light greenish, with some ferruginous streaks—scarcely at all cemented—having laminæ dipping north 45°, ... 4 ft.
E. Limestone, thin-bedded, with lenticular structure and undulating laminæ—streaked with dark, effloreсces with a salt having a cool and somewhat bitter taste, resembling epsom salts, ... 3 ft.
D. Sandstone, greenish-gray, friable—the lower half browner and harder, ... 2 ft.
C. Gypsum, massive, white, hard, in small masses impressed in the upper part of B. ....
B. Limestone, brown, glistening, hard, with streaks of green, ... 8 in.
A. Sandstone, bluish or greenish, moderately hard, with concretionary masses harder and more brown, ... 3 ft.

The dip seems to be from this point both north and south.

The strata, E, become in places highly ferruginous, and exhibit a tufaceous structure, which is probably a recent change in the rock. In other places it becomes a true breccia, with angular fragments of a brown limestone, held together by a tufaceous cement. Small stalactites are forming in places where the rock overhangs.

About 20 rods south from the main bluff, a blue clay is seen at the water's edge, in place of D, the strata, E, being commingled with the subsoil.

The layer, B, is not very persistent, being sometimes quite sandy, and passing into A.

The gypsum is in places imbedded in the blue sandstone A, in belts.

To the north of the main section, the limestones, E, become more arenaceous, and the sandstones, D, become shaly and increase to 4 feet, while the upper part of A is blue clay. The sand, F, becomes 5 feet. The amount of gypsum increases making a varying bed from six to twelve inches thick. A second gypsum layer appears in A, thicker than the other.

It is unlikely that a bluff of materials embracing so much soluble matter, has remained exposed to the action of the atmosphere and the lake, without undergoing important changes from its original character. Especially are we unable to decide from this exposure, as to the whole probable thickness of the gypseous deposites. The waves of Lake Huron have for ages been breaking against the exposed edges of the strata, and the gypsum has necessarily been dissolved out to a considerable distance back from the shore. In confirmation of this inference I found at Plaster Point, one mile north of the north line of Bay county, numerous "sink holes," as if produced by the subsidence of the overlying beds, after the dissolving out of the gypsum. Some of these are eight feet deep. Water is standing in them, probably at the level of the lake. The sides are steep, exactly as if the rocks had sunken. In one place a sink is seen pursuing an irregular course for several rods, toward the lake, and the whole appearance is exactly such as is produced by the falling in of the roof of a miner's "drift." Off this point, in calm water, the bed of the lake is seen to be a mass of pure white gypsum—the same, undoubtedly, which rises above the surface at Plaster Bluff half-a mile further north.

The land, back from the shore of Tawas Bay, rises in a succession of ridges running parallel with the lake. About 12 miles back, in the vicinity of the Au Gris river, the country becomes very broken, resembling that upon Grand River in the vicinity of the gypsum quarries. At the residence of Sherman Wheeler, 4½ miles south of Tawas City, one of the parallel ridges has attained an elevation of 40 feet, and the acclivity facing the lake presents a slope of 30° with the horizon. This ridge is said to increase in height as far north as the Tawas River. Mr. Wheeler informed me that no explorations had ever been made in this bluff, except to a limited extent by Mr Challis. Strata were found, called clay by Mr. Wheeler, though the specimen shown me was the brown limestone of the Michigan Salt Group, as seen at Plaster Bluff. The green streaks in it were pronounced by Mr. Challis to be indications of the proximity of copper. Coal was also prophesied in this ridge.

Arriving at Tawas City, I obtained some further information

from Mr. C. H. Whittemore. He says that a "slate rock" can be seen off White Stone Pt., extending out a mile from shore into 12 to 18 feet of water. He has traced this north to the neighborhood of Wheeler's (630), where it approaches within 8 or 10 feet of the surface, and disappears beneath the sand. Mr. Whittemore has bored 30 feet at Tawas City, to strike it, but thinks he has not succeeded. "It appears like a blue rock. Challis says it is iron ore." Mr. Whittemore bored 24 feet in the pure clay at the foot of the bluff, back of Wheeler's. He bored in several other places, including the top of the bluff, and found nothing but soft clay. It will be noticed that the statements of Mr. Whittemore are at variance with those of Mr. Wheeler. In the present state of the case, while it is obvious that gypsum occurs in considerable quantity along the shore of Tawas Bay, it is necessary that borings should be carefully made in several places, under the direction of a competent geologist—by which I do not mean one of those who search in the Michigan Salt Group for coal, iron and copper.

The rocks of the Salt Group should be found gradually rising toward the north along the shore of Tawas Bay. The gradual rise of the ridge, back of Wheeler's, conforms to this condition. If this is the case, the gypsum of Plaster Bluff and Whitestone Pt., should be found in this ridge. The limestone shown by Wheeler as taken from the ridge by Challis, belongs to one of the beds exposed at Plaster Bluff. In spite, therefore, of the negative results said to be obtained by Mr. Whittemore, I cannot resist the conviction that adequate explorations along this shore would be amply rewarded.

A short distance south of Tawas City, this ridge is cut through by Dead Creek, which has very high banks. Though no rocks are seen in place, numerous angular fragments are said to occur. In town 23, range 7, on the Ottawa river, are some hills 200 feet high, cut through by the river.

The region between the head of Tawas Bay and Kent county, has not yet been geologically explored. It is likely that numerous valuable facts could be gleaned from the notes of the linear surveyors. Mr. Wm. B. Hess, of East Saginaw, has in his custody maps and notes of resurveys, copies of which I had hoped to procure, but the limited means at my disposal have not been sufficient to enable me to incur the expense of copying them In the mean time there is little doubt that the rocks of the Michigan Salt Group outcrop in a broad belt arching northward from Tawas Bay through Ogemaw, Roscommon, Missaukee, Wexford, Lake and Newaygo counties. The interests of the State demand that explorations be made across this region at as early a day as possible. If the indications observed, be found verified, this resource will prove of incalculable value to the central counties of the peninsula, at present cut off from all ready communication with other parts of the State.

On the east side of Saginaw Bay some clays were seen at the mouth of Pigeon river, in Huron county, which very much resemble those of the Salt Group; and as this is about the place for the formation to strike the main land again, after crossing the bay, there are reasons for undertaking some more thorough explorations in that vicinity. Indications also exist of the occurrence of the formation in Tuscola county. It is likely, however, that the group thins out toward the south and nearly disappears through Lapeer, Oakland, Washtenaw, Jackson and Eaton counties, thus furnishing another illustration of the thickening of our formations toward the north. The salt springs at Saline, in Washtenaw county, and at various points in Jackson, may possibly issue from the attenuated representative of the group; but I am more inclined to think that these waters, like similar ones in Branch, Oakland, and the northern part of Huron county, are supplied by the various formations outcropping at these localities. Borings for salt have shown the Napoleon and Marshall sandstones to be saliferous, while at Saginaw, water from the coal measures stood at 1° of the Salometer in the upper part, and increased to 14° before reaching the Parma Sandstone. It is important to bear in mind that the occurrence of a brine spring proves nothing more than that there is salt somewhere in the State.

Comminuted carbonaceous matter is found in considerable abundance in some of the shales of the Salt Group. Besides this, no organic traces have been discovered.

14.—*Carboniferous Limestone.*

The best known outcrop of this formation is at Grand Rapids, in Kent county, where the Grand River experiences a fall of about 18 feet in the space of two miles. The rock here exhibits gentle undulations, but the resultant dip is slightly toward the north-east. It occurs in generally thin, irregular beds, which are considerably broken up, and embrace frequent partings of argillaceous and bituminous matter. In composition, it is generally eminently calcareous, but in the lower portion, arenaceous matter gradually gains preponderance; and belts and patches of the same material are irregularly distributed through the formation. In the upper part of the exposure here, is a belt, 5 feet thick, of red, ferruginous, arenaceous limestone. The thickness of the formation below this is 51 feet, while the thickness above, at this point, is unknown, though it is probably less than that below. The portion of the formation below the ferruginous stratum, contains numerous geodes, filled with brown and white dog-tooth spar, brown pearl spar, rhombic calcareous spar, selenite, anhydrite, aragonite, pyrites, &c.

From Grand Rapids; the formation has been traced north through Ada and Cannon, in Kent county, and to the rapids of the Muskegon, in Newaygo county. South of Grand Rapids, it is followed through Walker, Paris and Gaines, in Kent county, to Bellevue, in Eaton county, and thence by numerous outcrops to Parma, Sandstone, Spring Arbor, Summit, and Leoni, in Jackson county. The S. W. $\frac{1}{4}$ of S. E. $\frac{1}{4}$, sec. 13, Summit, is believed to be the most southern well-characterized exposure of this formation. It occurs in a quarry belonging to Michael Shoemaker. The section exposed here is about 14 feet, as follows:

D. Sandstone, red, calcareous, highly shattered, breaking into cuneiform fragments, with conchoidal surfaces, changing locally to C.,........................ 5 ft.
C. Limestone, highly ferruginous, brecciated in places, containing nodules of chert. Passes upward into D. 4 ft.
B. Limestone, quite arenaceous, brecciated, shattered, with thin layers which are sandy and greenish; the whole exterior of some of the blocks covered with a thick, loose coating of the same material; upper surface undulating, but smoothed as if by aqueous action, before the superior layers were deposited,... 2 ft.
A. Limestone, compact, crystalline, silicious, bluish-gray, with some crystals of dog-tooth spar. Exposed,... 3 ft.

The stratum D. is the parting layer between the upper and lower portions of the formation. The characters of this bed are exceedingly uniform at all the outcrops on the south and west sides of the geological basin.

At the quarry of C. Roberts, S. E. ¼ N. W. ¼ sec. 17, Spring Arbor, is found a section similar to the preceding:

D. Sandstone, ferruginous, highly calcareous, breaking with cunoidal fracture.
C. Limestone, bluish-gray, hard crystalline, thick-bedded, of excellent quality, containing small crystal-lined geodes,...................................... 5 ft
B. Limestone, ragged, arenaceous, with irregular seams and blotches of greenish sandstone,............... 2 ft.
A. Limestone, fine-grained, hard, bluish,............... 2 ft.

This formation outcrops on sections 21, 26, 27, 28, 29, 31 and 32, in the township of Bellevue, Eaton county. From the various quarries in the vicinity of the village, the following succession of strata was made out:

G. Limestone, thick-bedded, calcareous,............... 3 ft.
F. Limestone, yellow, silicious,....................... 2 ft.
E. Limestone, massive, destitute of fossils,............ 6 ft.
D. Belt filled with a cæspitose Cyathophylloid,......... 6 in.
C. Limestone, thick-bedded, containing *Allorisma* and a large coiled shell (not seen).
B. Limestone, blue, compact, hard, thick-bedded, containing geodes.
A. Sandstone, bluish-gray.

From Leoni, in Jackson county, no actual outcrop of this

formation is known, until reaching Tuscola county. On the S. E. ¼ sec. 16, T. 12 N., 9 E., are found numerous fragments of a compact, blue, non-fossiliferous limestone, which has been quite extensively employed for burning. Rock is also felt in the bed of Cass River, at this place.*

On the S. W. ¼ sec. 22, T. 16 N., 9 E. (Sebewaing), Tuscola county, are found abundant fragments of the lower arenaceous member of the Carboniferous limestone, containing an *Allorisma*. Further north, on the N. W. ¼ sec. 13, T. 16 N., 9 E., half a mile above the mouth of Cheboyong Creek, is a distinct outcrop of an arenaceous, yellowish limestone, containing numerous specimens of *Allorisma clavata*, and other species identical with the one found at Grand Rapids and Bellevue. In this sandstone occurs a thin layer, highly calcareous and exceedingly tough. The next outcrop occurs at the northern extremity of Stone Island (Shung-woi-gue), in Saginaw Bay. The rock rises but four or five feet above the water, and is generally much brecciated. It is for the most part calcareous, but exhibits bands and patches of an arenaceous character; and the beach is strewn with fragments apparently thrown up from a greater depth, which seem to belong to the arenaceous strata exposed at Cheboyong Creek. *Allorisma* occurs here, and geodes are not unfrequent. The lower arenaceous layers appear again on North Island (Ash-qua-guin-dai-gue).

On the south side of Wild Fowl Bay, is a characteristic outcrop of the formation, extending along nearly the whole shore. The dip is very slight toward the south-east. The greatest actual exposure is only four feet, but the rock undoubtedly rises in the bank to the height of 15 feet above the water. The following stratigraphical characters were noted:

F. Limestone, argillaceous, cherty, perforated extensively by a *Syringopora*, ... 10 in.
E. Limestone, compact, bluish, weathering white, ... 1 ft.

*About the forks of the Cass, above and below, are found numerous fragments of a limestone of quite different character, and some kind of rock is felt with a pole in the bed of the stream. The limestone is dark argillaceous, and occasionally arenaceous. This is the *pipestone* from which the Indians of this vicinity cut their pipes. It has somewhat the appearance of a hydraulic limestone.

D Limestone, arenaceous, with nodules of chert. Seen dipping into the water 10 rods west,.............. 4 in.
C. Limestone, dark, calcareous, with bituminous (flinty?) streaks and laminæ—intersected by broad cracks which have been subsequently filled with material like D,..10 in.
B. Limestone, yellowish, highly arenaceous, thin bedded, rather incoherent, the lower one-fourth curiously banded with lighter and darker streaks,................$1\frac{1}{2}$ ft.
A. Limestone, arenaceous, highly shattered and recemented.

The flint nodules in the layer D, are bluish, of a fine, homogeneous structure and strike fire with steel, with great readiness. They exist in large quantity. Should there be a demand for such an article, Wild Fowl Bay could furnish an abundant supply.

The layer E, would furnish a superior building material.

The layers D, E and F, will make excellent lime, and the elevation above the water, especially if the rock enters into the formation of the high bank along here, would fully justify the opening of a quarry.

From this point the formation crosses Saginaw Bay, and next appears on the Charity Islands. The rock is seen under water for a long distance south-west of Little Charity Island. It outcrops along the northern, western and southern shores, consisting of one or two layers 12 to 15 inches thick. It abounds in the *Syringopora*, before referred to, and contains some concretions of a cherty nature. It is replete with *traces* of organic remains, but nothing is well preserved or identifiable, save some *Bryozoa* and *Cyathophyllidæ*.

The formation outcrops more extensively on the north side of Great Charity Island, where it rises about five feet above the water, and presents the following section:

C. Limestone, areno-calcareous, containing *Bryozoa*, *Cyathophyllidæ* and *Allorisma*,..........................10 in.
B. Limestone, with cherty nodules,.....................10 in.
A. Sandstone, calcareous, obliquely laminated,......... 4 ft.

Some portions of A are well characterized sandstone, of a

whitish or grayish color. The laminæ extend from top to bottom of the mass, dipping north-west at an angle of about 45° They are quite undulating and even contorted, and the whole mass shows something of a rude concretionary structure.

The formation strikes the main land at Point au Gres. The rock here, in spite of the name, is a limestone. That part of the outcrop above the water consists of three layers, each about 15 inches thick. The upper layer is, in places, quite arenaceous, but is more solid than the rock at Cheboyong creek. It contains stains of greenish matter and irregularly cylindrical, somewhat concretionary, bodies, considerably colored. Here occurs the *Syringopora* which occupies the top of the section at Wild Fowl Bay. This layer is separated from the next by two or three inches of laminated, argillo-calcareous sandstone.

The middle layer is more purely calcareous, but contains some sand. Here I saw an *Acervularia*, a *Syringopora*, a *Cyathophylloid*, and the remains of a bony body, whose impression left pits regularly disposed upon the rock.

The lower layer abounds in concretionary cherty nodules, perforating the rock in every direction, often appearing, when broken at the surface of the stratum, like plugs driven into the rock. These nodules are less flinty than those seen at Wild Fowl Bay.

A large *Productus* was picked up on the beach.

Between this point and Newaygo county, no definite information has yet been collected respecting this formation. We know from the surveyors' notes, that limestone outcrops at various points, but I have seen no specimens, and its geological characters are in doubt.

The thickness of this formation is much greater on the western (and probably northern) borders of the basin, than on the southern. It is 51 feet thick at Scribner's well at Grand Rapids, and the whole thickness in this vicinity is probably not less than 70 feet. It is found 65 feet thick in the salt wells of East Saginaw.

As this calcareous member of the Carboniferous system possesses great economical and scientific interest, I have thought best to enter into a greater amount of local and stratigraphical detail than I have done in respect to the other groups. For the determination of the parallelism between this formation and the carboniferous limestones of the North-west, lithological considerations become the more important, from the great scarcity of fossils in our formation, and the entire absence of those forms which furnish the means of certain identification in Indiana, Illinois, Missouri and other States.

From the account which has been given, there is obvous difficulty in identifying our limestone with any of the groups that have been established by the researches of Owen, Hall, Swallow, Worthen, McChesney and others. Little attention has yet been given to fossil remains, but the following notes of species thus far observed, may be here recorded:

Notes on the Fossils of the Carboniferous Limestone of Michigan.

[The numbers prefixed refer to the University Catalogue.]

POLYPI.

237. Lithostrotion mammillare, *Edwards and Haime.*
The specimens agree entirely with figures and descriptions by Hall (Iowa Rep.) and Owen (Geol. Iowa, Wisconsin, &c.).

Localities—Grand Rapids and boulders in that vicinity.

250. Lithostrotion (Lithodendron) longiconicum? *Phillips.*
This abundant, generally distributed, luxuriantly cæspitose and branching Cyathophylloid, presents externally the non-striated appearance of *L. longiconicum*, while it has the oval columella of *L. sociale*, Phillips. It is less straight than either of these species, and not improbably constitutes a distinct type.

Localities—Grand Rapids, Bellevue, Great Charity Island, Pt. au Gres.

252. Cyathophyllum fungites, *De Koninck.*
These specimens considerably resemble *Turbinolia fungites*, Fleming, (Phillips, Geol. Yorkshire, Pl. III, Fig. 23.) They are less broad than the figure of De Koninck, (Animaux Foss. de Belg. Pl. D, Fig. 2,) but agree well with Owen's figure of the same, (Iowa, &c., Table IV, Fig. 4.) The last named is reported from the Keokuk rapids.

Localities—Grand Rapids, Stone Island, (Saginaw Bay.)

253. Cyathophyllum, sp?
More expanded than the preceding, and more irregular in its outline.

Locality—Grand Rapids.

251. Caryophyllia duplicata, *Martin.*
Agreement very good.

Locality—Great Charity Island.

249. Acervularia, sp?
The obscure styliform elevation in the cup of this large coral strikes the eye at first as belonging to a *Lithostrotion*, but after careful examination, I am convinced that the coral possesses no columella As far as its characters can be inspected, it does not differ from *Strombodes*, as restricted by Pictet. It has the general aspect, however, of an *Acervularia*, and only differs in having the transverse floors more numerous in the visceral chamber.

Locality—Pt. au Gres.

248. Syringopora, sp?
Tubes small; much geniculated and with numerous oblique connecting tubes or bars as large as the main tubes. A very characteristic fossil, but very obscure.

Localities—Grand Rapids, Pt. au Gres, Wild Fowl Bay, Great and Little Charity Islands.

ECHINODERMATA.

236 Archæocidaris.
Remains of spines only, which more resemble Hall's figures of *A. Agassizii,*[*] (Burlington Limestone,) and *A. Keokuk*, (Keokuk Limestone,) than any others accessible to me.

Locality—Grand Rapids.

BRYOZOA.

238. Fenestella membranacea (?), *Phillips.*
Very closely related to *Gorgonia (Retepora) membranacea*, Phillips and DeKoninck. The fenestrules, however, are but little longer than broad.

Locality—Grand Rapids.

339. Fenestella, sp.?
Similar to the preceding, but the fenestrules are more elongated and less quadangular. A distinct, sharp keel runs along the ray between the two rows of cellules. The form and disposition of the cellules is a miniature representation of

the fenestrules. This species is scarcely distinguishable from specimens collected from the St. Louis limestone, two and a-half miles west from Charboniere, on the Missouri River. It must bear considerable resemblance to *F. patula,* McCoy.

Locality—Grand Rapids.

241. Fenestella, sp.?
The rays are very narrow and flexuous between the lines of small roundish fenestrules. No cellules have been seen.

Locality—Grand Rapids.

240. Polypora, sp.?
Allied to *P. Shumardii,* Prout, (Trans. Acad. St. Louis, I., p. 271), a Devonian species. The cellules are exceedingly minute and indistinct, though I imagine the specimen shows the reverse side

Locality—Grand Rapids.

242. Cladopora (?) sp?
Prof. Hall's characterization of this genus, (Pal. of N. Y., II, 137,) does not disagree with these specimens.

Locality—Grand Rapids.

243. Cladopora, sp?
The cellules are promiscuously arranged upon a flattened, branching axis.

Locality—Great Charity Island.

244. Coscinium (?)
Fenestrules quincuncially disposed on a flattened branching axis, which is minutely porous.

Locality—Great Charity Island.

245. Monticulipora (?) sp? (Rhinopora, *Hall.*)
Minute, polygonal, crowded, rayless pores or cells, irregularly studding the surface of a compressed, lobated, subspheroidal mass.

Locality—Grand Rapids.

247. Ptilodyctia (?) sp? (Stictopora, *Hall.*)
A branched, nearly terete stem, with pores apparently on all sides.

Locality—Grand Rapids.

BRACHIOPODA.

207. Productus Altonensis, *Norwood and Pratten,* (Journal Acad. Nat. Sci., Phil. [2| III, 7.)
Agrees very well. It bears some resemblance to *P. costatus* Sowerby, *Var,* Hall, (Iowa Rep. p. 712,) but it is hardly

broad enough for this variety, and has no mesial sinus though apparently flattened. It also resembles *P. costatus*, Sow. (DeKoninck, Carbon. Foss. Belg. p. 164,) but it is not broad enough nor does it exhibit any granulations. The striation of some specimens is exceedingly like that of *P. comoides*, Sow. (Phillips, Geol. Yorkshire, Pl. VII, Fig. 4,) *P. Altonensis* is from the upper portion of the Carboniferous limestone, (St. Louis Limestone of Hall.)

Locality—Grand Rapids.

208. Productus pileiformis, *McChesney* (New Pal. Foss. p. 40). It bears some resemblance to *P. ovatus*, Hall, but the striæ are too fine. It differs in the same way from *P. Altonensis*, and is, besides, broader than that species. From *P. tenuicostus*, Hall, it differs from the mode of increase of the striæ which is by intercalation instead of bifurcation. The concentric rugæ, moreover, are not sufficiently conspicuous. *P. pileiformis* is from the Kaskaskia Limestone.

Locality—Ferris' limekiln, 2 miles north of Jackson.

209. Productus.
This was thought by Prof. McChesney, (without making direct comparison) to be his *P. fasciculatus* (Op. cit. p. 38). It does not, however, agree sufficiently well.

It has some resemblance to *P. Cora*, D'Orbigny, as figured by Owen (Op. cit. Table IV., Fig. 1), but my specimen exhibits seven or eight ribs raised higher than the intervening ones. The surface is covered by fine undulating concentric striæ, and when exfoliated, shows numerous punctate depressions beneath. The spine bases are very scattered on the ventral valve.

Locality—Unknown. Supposed to be with preceding.

210. Productus Wilberanus (?), *McChesney*.
The agreement is moderately good, but the concentric striæ or rugæ are not generally distributed. This species comes from the coal measures of Charboniere, Mo. It differs from *P. scabriusculus*, DeKoninck, (Op cit., Pl. IX., Fig. 5,) by the irregular disposition of the pustules.

Locality—Unknown. Supposed to be with the preceding.

211. Productus, sp?
A delicate species much too finely striated for *P elegans*, Norwood and Pratten, and somewhat too much so for *P. Altonensis*, while it is too regularly and coarsely striate for *P. pileiformis*.

Locality—Stone Island, in Saginaw Bay.

212. Productus, sp?
A species destitute of radiating striæ or ribs. The ventral valve shows five or six varices or lines of growth; the dorsal, (if it is the same species) more than this. The surface is punctate.
It is allied to *P. gryphoides* DeKoninck, (Op. cit. p. 182, Pl. IX., Fig. 1, but not to the other figures.)
Locality—Grand Rapids.

213. Productus, sp?
Considerably resembles in form and size, *P. muricatus*, Norwood and Pratten, from the coal measures, but the radiating ribs are too fine. It is somewhat like DeKoninck's figure of *P. costatus*, (Pl VIII., Fig. 3,) but is not sinuated. It closely corresponds with *P. costatus* from tbe coal measures, 9 miles north of St. Louis, Mo., except in not being sinuated, and in the less conspicuous character of the concentric rugæ.
Locality—Grand Rapids.

214. Productus, sp?
Has the form of the last, but the radiating striæ are much finer and the concentric folds little conspicuous.
Locality—Grand Rapids.

215. Orthis umbraculum (?) *von Buch.*
Allied to *O. robusta*, Hall, (Op. cit. 713.) Resembles *O. umbraculum* as figured by DeKoninck, (Op. cit. p. 223,) from carboniferous limestone, but better as figured by Owen, (Op. cit., Tab. V., Fig. 11.) It exceedingly resembles this species as figured by Hall in Stansbury's Report, (Pl. III, Fig. 6.)
Locality—Grand Rapids.

216. Orthis (?) sp?
A single flat valve with slender auriculate appendages extending the hinge line to nearly the greatest width of the shell.
Locality—Grand Rapids.

232. Orthis, sp?
A single dorsal valve more finely striate than the preceding.
Locality—Stone Island, in Saginaw Bay.

217. Athyris subquadrata, *Hall*, (Iowa Rep., 703.)
This species is from the Kaskaskia limestone.
Locality—Grand Rapids.

218. Athyris sublamellosa, *Hall*, (Iowa Rep., p. 702.)
Agrees pretty well, but the shell is not more than one-third the dimensions of Hall's, and is proportionally a little thinner. This species is from the Kaskaskia limestone.

Locality—Unknown. Supposed to be with the last.

219. Athyris, sp ?
Considerably resembles *Terebratula rhomboidea.* Phillips. Also has some affinities with McChesney's *A. obmaxima* from the Keokuk limestone

Locality—Grand Rapids.

220. Athyris Roysii (?) *DeKoninck.*
Closely related to smooth specimens of this species. (Op. cit, Pl. XX, Fig. 1)

Locality—Grand Rapids.

221. Athyris, sp ?

Locality—Grand Rapids.

222. Athyris, sp ?

Locality—Grand Rapids.

223. Athyris, sp ?

Locality—Grand Rapids.

225. Athyris ? sp ?
Bears considerable relationship to *Terebratula Roysii, var,* Leveille, (DeKoninck, Pl. XXI, Fig. 1,) but it is somewhat broader and more quadrate. It resembles *A. obvia* McChesney, (p. 81,) but differs in having its two valves equally convex, and in its faint radiating lines. It is less sinuate than *A. differentius,* McChesney.

Localities—Grand Rapids, Pt. au Gres.

229. Athyris ? sp ?

Locality—Ferris' limekiln, 2 miles north of Jackson.

224. Terebratula sacculus, var. hastata (?) *DeKoninck.*

Locality—Grand Rapids.

226 Terebratula subretziæforma (?) *McChesney.*
It is broader proportionally than this species from the Kaskaskia limestone, but otherwise it agrees closely. The correspondence is also very good with *T. subtilita,* Hall, (Stansbury's Rep., p. 409,) but our shell is smaller. It is smaller, thinner and less deeply sinuated than specimens of *T. subtilita,* Hall, from the coal measures of La Salle, Ill.

Locality—Grand Rapids.

227. Terebratula, sp ?

Locality—Grand Rapids.

228. Terebratula, sp ?

Locality—Grand Rapids.

233. Terebratula, sp?

Locality—Stone Island in Saginaw Bay.

235. Atrypa, sp?

Large, very gibbous, with numerous radiating ribs.

Locality—Pt. au Gres.

234. Spirifer Keokuk, var, *Hall.*

The general aspect is of this variety. The principal points are as follows: The valves are nearly equally convex; the mesial sinus of the dorsal valve is abruptly elevated, but instead of being divided into four distinct plications, presents but two, which are distinctly marked. The ventral valve has a sinus, simple at its origin, but soon divided by a distinct plication, and in some specimens by two lateral ones. Surface on each side of the mesial fold and sinus marked by about 8 plications, the two adjacent to the mesial sinus and elevation originating from a single one near the apex. No striæ are seen on the surface.

The variety above is from the St. Louis limestone.

Our shell bears also a close relation to *S. rotundatus* var. *planata*, DeKoninck (Pl. XIV, Fig. 2, and XVII, Fig. 4) It is, however, a little more transverse than these figures, approaching in outline *S. acuticostatus*, DeKoninck.

Locality—Grand Rapids.

LAMELLIBRANCHIATA.

201. Allorisma clavata, *McChesney.*

The "flattened or slightly concave space extending like a sinus from the beak to the base" of the shell is scarcely perceptible in my specimens. From the Kaskaskia Limestone.

Locality—Cheboyong Creek, Tuscola county.

202. Allorisma, sp?

Bears some resemblance to *A. sinuata*, McChesney, from the Kaskaskia Limestone, but differs thus: Not prolonged posteriorly, nor extremely gibbous; beaks less than one third the length of the shell back from the anterior extremity; ventral margin but very slightly sinuate, and surfaces of valves scarcely depressed; concentric ridges not very strong

In form and makings it resembles *Lithodomus Jenkinsoni*, McCoy, (Brit. Pal. Rocks and Foss. Pl. 3 F. Fig. 2).

Localities—Grand Rapids, Stone Island, Great Charity Island.

203. Allorisma, sp?

Allied to *A. sulcata*, Fleming (King, Permian Fossils, Pl. XX. Fig 5), "from carboniferous shales, Redesdale, Northumberland." I cannot, however, detect any radiating pimples. This form also resembles *A. regularis* (Owen, Rep. Iowa, &c., Tab. V. Fig. 13).

Localities—Grand Rapids, Cheboyong Creek.

204. Allorisma, sp?
The umbones almost overhang the anterior extremity.

Localities—Grand Rapids, Cheboyong Creek.

206. Allorisma, sp?
Very gibbous, umbones sharp, incurved, overhanging the anterior extremity.

Localities—Grand Rapids, Cheboyong Creek.

205. Nucula? sp?
Perhaps an *Allorisma* or *Myalina;* very obscure.

Locality—Grand Rapids.

231. Nucula? sp?

Locality—Cheboyong Creek.

230. Myalina lamellosa (?), *DeKoninck.*
The beak is rather too much reourved, and the surface too smooth, for this species. It has some resemblance to *M. Swallovi*, McChesney (Op. cit., p. 57), from the coal measures.

Locality—Grand Rapids.

GASTEROPODA.

Remains of Euomphaloid casts have been found at Grand Rapids. Very complete coiled shells have been met with at Bellevue, in Eaton county, but I have not been fortunate enough to secure any.

TRILOBITES.

254. Phillipsia, sp?
Fragments of tails, resembling *P. Brogniarti*, Fischer (De Kon. Op. cit., Pl. LIII, Fig. 7).

Locality—Grand Rapids.

255. Phillipsia, sp?
Fragments of two tuberculated tails.

Localities—Grand Rapids. From Great Charity Island is a portion of a head which may belong to the same species.

FISHES.

Remains of spines and Psammodus-like teeth have been met with at various localities.

Of the fifty-five species discriminated in the foregoing list, only sixteen have been even doubtfully identified with species hitherto described. Of these sixteen, twelve only are known to the writer to occur in the carboniferous rocks of the Western States, and are distributed as follows:

In the Coal Measures: *Productus Wilberanus (?).*

In the Kaskaskia Limestone: *Productus pileiformis, Athyris subquadrata, A sublamellosa Terebratula subretziæforma (?) Allorisma clavata.*

In the St. Louis Limestone: *Lithostrotion mammillare, Productus Altonensis, Spirifer Keokuk, var.* Also the *Fenestella*, No. 239.

In the Keokuk Limestone: *Cyathophyllum fungites (?)*

In the Carboniferous Limestone generally: *Orthis umbraculum.*

In the present state of our knowledge, it would be premature to attempt to identify the Carboniferous Limestone of Michigan with any of the group of Illinois and neighboring States. Attention may, however, be directed to the following points:

1. No indications of *Archimedes* have been detected in the formation.

2. Very few indications of Encrinites have been discovered.

3. The probable identification of five species, with forms belonging to the Kaskaskia Limestone, affords a pretty strong indication that at least some parts of our formation lie in the horizon of the very top of the general series.

4. The identification of four species with forms from the St. Louis Limestone, foreshadows a strong affinity with that part of the system. The brecciated character of many portions of the rock, points, if anywhere, to the same relationship.

5. The ferruginous, arenaceous stratum, occurring in the midst of the formation, may easily mark the boundary line between the two successive epochs last mentioned; although, at present, it is impossible to say whether the distribution of the fossils conforms with such a separation.

6. The arenaceous character of the lower part of the formation, becoming on the Charity Islands, a thick mass of yellowish sandstone; the blotches and disseminated particles of greenish matter found here; the frequent shaly partings of the strata; certain "vermicular ramifications" upon the bedding surfaces, all recall the characters of the upper part of the Warsaw Limestone.

7. At the same time, the portion below the ferruginous arenaceous bed abounds in geodes filled with crystals of calc spar, dog-tooth spar, pearl spar, selenite, anhydrite, pyrites, &c., which recall the "Geode Bed," below the Warsaw Limestone.

Whether our formation possesses real affinities with all the groups from the "geode bed" to the Kaskaskia limestone, is an interesting question which probably we shall yet be able to resolve. Such a result would not be surprising. The various groups of the Silurian and Devonian Systems, stretching through a vertical thickness of many thousand feet in New York and Pennsylvania, are all faithfully represented within the space of as many hundred feet in the Lower Peninsula of Michigan; and we are prepared to foresee that our situation, similarly, upon the borders of the great carboniferous sea, has resulted in an attenuated representation of the various groups of the carboniferous limestone, which towards the south-west thickens up to some thousands of feet.

15.—*Parma Sandstone.*

In the townships of Parma, Sandstone and Springport, in Jackson county, is found a white, or slightly yellowish, quartzose, glistening sandstone, containing occasional traces of terrestrial vegetation. On the line between sections 18 and 19, in the township of Sandstone, this rock is seen succeeding upwards to the furruginous bed of the Carboniferous limestone. On the N. W. ¼ of N. W. ¼, sec. 29, at the quarry of Mrs. Titus, the sandstone presents a characteristic exposure. The rock is light-colored, thick-bedded, firmly cemented and appears to furnish an excellent article for building purposes. It presents the remarkable dip of 45° SSW, with vertical divisional planes running parallel with the strike. The rock is occasionally stained with iron, is of medium fineness and glistens in the sun, from the glassy clearness of the quartzose grains. For caps and sills it is apparently superior to the Napoleon sandstone.

This quarry occurs upon a ridge elevated about 35 feet above the limestone, which is exposed over an area of a square mile, beginning a few rods further west. It has every appearance of a violent uplift, but the undisturbed position of the underlying limestone seems incompatible with this supposition, and we are forced to conclude that the apparent dip of the formation is nothing more than a very illusory example of oblique lamination.

At the N. E. ¼ of S. W. ¼, sec. 18, Parma, near where the highway crosses Rice creek, this sandstone affords a *Calamite*. At the S. W. ¼ of N. E. ¼, sec. 19, Parma, it has been quarried by Mr. L. H. Fisk. The rock is nearly white, sometimes varying to a light straw color; and in some places is quite full of small white quartzose pebbles. A portion of the Albion flour mill was built of stone from this quarry.

Very numerous quarries have been opened in this formation in the northern part of Jackson county, but it is unnecessary to particularize at present.

From Mrs. Titus' quarry, the outcrop trends south-east toward the village of Barry, and is worked at several points. At Boynton's quarry, half a mile north-west of the Barry coal mines, is a fine exposure of massive sandstone, which, though occupying a higher geographical position than the coal, is nevertheless believed to belong geologically below it. It is found above the limestone in the vicinity of the quarry of Chester Wall, and seems to be the highest rock throughout most of the interval between Barry and Woodville coal mine. South of Woodville it may be recognized by its glistening character, to the immediate vicinity of Hayden's coal mine, and from here to the region south of Jackson. It is unnecessary to particularize localities. Indeed, it is separated in this part of the State, by so short an interval, from the Napoleon Sandstone below, and the Woodville Sandstone above, that the geographical distribution of this formation has not been very accurately determined, even after a pretty careful survey.

This sandstone was pierced in the boring for salt at East Saginaw, and its thickness was found to be 105 feet. It cannot be a third of this on the southern border of the basin.

No fossils, except imperfect *Calamites* and vegetable traces, have been detected in the Parma Sandstone, but accompanying its outcrop, throughout its whole extent, are found angular fragments of a flinty or cherty sandstone abounding in impressions of *Sigillariæ*. Not unlikely these remains belong to the formation in question. They frequently recall the characters of the Ohio Buhrstone.

The Parma Sandstone occupies the geological position of the Ohio Conglomerate. The occurrence of pebbles at a single locality observed, constitutes a faint physical resemblance, but in other respects the correspondence is rather remote. Unlike the Ohio Conglomerate, it is separated from the upper Devonian rocks by a considerable thickness of calcareous and arenaceous strata.

16.—*Coal Measures.*

The Coal Measures, with the overlying Woodville Sandstone, occupy the whole central area of the Lower Peninsula. The territory covered, embraces the counties of Saginaw, Shiawassee, Clinton, Ionia, Montcalm, Gratiot, Isabella and Midland, and the greater part of Tuscola, Genesee, Ingham, Eaton and Bay, being nearly thirteen counties, besides considerable portions of Livingston, Jackson, and probably other counties on the north. The whole area underlain by the coal measures is approximately 187 townships, or 6,700 square miles. Over nearly the whole of this extent of country the measures will be found productive.

The southern border of the basin reaches probably into the township of Blackman, in Jackson county. Beyond this seem to be several detached outliers in which the measures do not attain their normal thickness, though the principal seam of coal is very little diminished. The most southerly point at which coal has been found in place, is at Hayden's mine, where it was discovered in 1835, in digging the foundation of a mill. This is on section 1, in the township of Spring Arbor. The opening occurs on Sandstone creek where it is crossed by the highway, on the $\frac{1}{8}$th section line running south through the S. E. $\frac{1}{4}$. The outlier seems to be embraced in a gentle elevation, covering, perhaps, 40 acres to the west of the opening. Some distance up the hill slope, a boring was made with the following result:

E. Drift materials, ... 8 ft.
D. Shale, ...22 ft.
C. Coal, ... 4 ft.
B. Under clay, ...14 ft.
A. Parma Sandstone.

In the drift which has been carried into the hill the coal is found only three feet thick, and contains a seam of Iron Pyrites one foot from the top. Fragments of Black Band Iron Ore are brought out, which contain impressions of fishes. The sandstone (A) comes to the surface a few rods to the north, and a boring for coal was executed in it, of course without success. The boring, however, became an Artesian well.

One mile north of Hayden's mine, occurs the Woodville mine, owned by the Detroit and Jackson Coal and Mining Company. We here find the coal measures overlain by a sandstone, which, from its good exposure in the cut of the side track from the mine to the Central Railroad, has been designated provisionally the Woodville Sandstone. The section passed in the shaft of this mine is as follows:

E. Superficial materials, ...12 ft.
D. Woodville sandstone, ...30 ft
C. Shales, dark bituminous, with 6 feet of fine light colored clay, ...43 ft.
B. Bituminous coal, ... 4 ft.
A. Under clays, ... 3 ft.

A few rods from the shaft toward the north-west, the sandstone was found 45 feet thick in a boring. The cut of the side track shows this rock to be strikingly marked by lines of oblique lamination, which generally dip toward the south. The rock has a pale buff color, unequally distributed, and is but moderately coherent, rather friable, and towards the top is wholly disintegrated.

The shales are compact, fine, black and highly bituminous. In traversing the drifts or chambers leading from the shaft, the shaly roof is seen to be somewhat undulating and to present many evidences of slight disturbance since solidification. It is intersected by numerous fractures, and in many instances the movements of the opposing faces against each other have polished them most perfectly. The blackness and solidity of the shale give specimens the appearance of polished jet. The shale contains a *Lingula* probably unknown to science.

The coal is bituminous, solid, generally free from foreign matters, but is intersected by a thin belt of iron pyrites which is also slightly disseminated through the contiguous portions of the coal. The coal furnishes a glistening coke, samples of which were much admired at the State Fair.

It is uncertain whether this outcrop is connected with the main basin or is only another outlier. Numerous explorations for coal have been made in vain on the N. E. ¼ sec. 36, Sandstone, and extending over the line into the N. W. ¼ of the section. At one point a boring was carried to the depth of 279 feet. The coal measures seem to be mainly denuded along the interval between Woodville and Barry. At the deep boring, the Parma Sandstone was found 24 feet thick; a series of calcareo-arenaceous strata holding the place of the carboniferous limestone, 22 feet; a series of argillo-arenaceous strata occupying the place of the gypseous, or Michigan Salt Group, 49 feet; the Napoleon Group, including 20 feet of separating shale at bottom, 114 feet. The boring extended 56 feet into the Marshall Group. With such an interpretation of the results of this experiment, it would be obviously inconsistent to encourage further expenditures in the exploration of rocks below the shales of the coal measures.

A little further west, at the village of Barry, the coal is found again outcropping and has been worked to a considerable extent by the Jackson City Coal Company. The coal possesses nearly the same qualities here as at Woodville, though what I saw seemed to be of a less solid character. Some specimens were furnished me, however, for exhibition at the State Fair, which, in physical characters, appeared equal to any in the State.

The geology of this vicinity is exceedingly complicated—the carboniferous limestone and overlying Parma Sandstone appearing at points north-east and north-west of the mines, at a higher geographical position than the coal. There can be no doubt that these mines are also situated on an outlier of the coal basin, of but limited extent

An outcrop of coal is said to occur about half a mile west of the village of Barry. East of here the coal is seen again outcropping in the bank of the Grand River at the mill-dam in the city of Jackson, and indications of its approach to the surface are seen at several other places in the neighborhood. Borings and excavations have been made at various points, with no uniform results. In the shaft which was sunk by the Jackson City Coal Company, the following section was passed, according to the statement of Mr. William Walker:

G. Superficial materials,............................ 3 ft.
F. Sandstone, white or slightly stained, banded below with ferruginous and argillaceous streaks; contains *Calamites* and carbonaceous matter,....................26 ft.
E. Black bituminous shale with *Lingula*,................14 ft.
D. Black band iron ore, with abundance of *Lingula*,..... 3 ft.
C. Cannel coal,.................................... 2 ft.
B. Bituminous coal,................................ 2 ft.
A. Finely arenaceous fire clay, with abundance of *Stigmariæ*,.. 7 ft.

In the boring close by, the section continues downward through 30 feet of arenaceous materials, probably representing the Parma Sandstone.

Numerous explorations have been made in the vicinity of the

city of Jackson, but it would occupy too much space to detail the results. It must here suffice to say in general terms, that the statistics accumulated seem at first view to constitute a perfectly chaotic mass, without the least trace of a fixed order of succession among the strata, but that after correcting the errors in the mineralogical language of the well borers, it is found that the different explorations have pierced the Woodville Sandstone, the Coal Measures and the Parma Sandstone; that these three formations present numerous sudden flexures, so that after denudation of the ridges, each has been brought to the surface at numerous points. The consequence is, that in some cases the exploration has commenced in the Woodville Sandstone, in others, in the Coal Measures, and in still others, in the Parma Sandstone, which is shown to consist in the lower part, of an alternation of quartzose and argillaceous beds. It further appears that the Artesian waters of this vicinity proceed from the lower portion of the Parma Sandstone, and that the trough shaped attitudes into which the rocks have been thrown, are exceedingly favorable to the reception and retention of large quantities of water. In a full report upon our geology, it will be interesting to exhibit the correspondence of the results of the various explorations and to illustrate the whole to the eye by appropriate diagrams.

As the three localities already referred to as the seat of coal mining operations are thought to be situated upon outliers of the great coal basin, so nothing more can at present be said of the city of Jackson—the indications being, that rocks below the coal measures occupy the surface to the north of the city.

Between Ingham and Genesee counties the boundary of the coal formation has not been traced. In the south-western part of the township of Mt. Morris and contiguous portions of Flushing, in the latter county, according to the observations of Dr. Miles, the shales and sandstones of the coal measures make numerous outcrops. On the S. E. ¼ sec. 26, Flushing, the following section is observed in the bank of the Flint River:

Superficial materials, ... 4 ft.
Black shale, containing *Lingula*, *Chonetes Smithii*, *Productus asperus* and *Spirifer cameratus*, ... 3 ft.
Sandstone, tinged with iron, ... 7 in.
Shale, ... 1 ft.
Sandstone, ... 3 in.
Shale, ... 4 ft.
Sandstone, ... 6 in.
Shale to surface of water, ... 10 in.

A short distance west of here the section is seen to be extended upwards by the superposition of 7 inches of sandstone and five feet of an overlying shale. The bed of the river here is covered by a somewhat undulating and shattered gray sandstone which is considerably quarried for building. At a point on the N. E. ¼, section 35, Flushing, a sandstone was seen to attain a thickness of about 12 feet, in an excavation made by Mr. Niles.

On the N. E. ¼ of S. W. ¼, section 22, Flushing, a shaft was sunk on the farm of A. J. Brown, of which the following account was obtained:

Superficial materials, ... 14 ft.
Sandstone, below, bluish, gritty, ... 8 ft.
Coal, ... 2½ in.
"Horseback claystone," (Blackband?) .... 2 ft.
Same with kidney iron ore, ... 2 ft.
Shale, ... 5 ft.
Sandstone and *salt water*, ... 3 ft.
Shale, ... 4 ft.
"Black hard stone," combustible, ... 4 ft.
White Fire-clay, ... 2 ft.
Hard white sandstone, ... 2 ft.
Darker Sandstone, ... Unknown.
Striped sandstone, ... 3 ft.
Shale, ... Unknown.
"Coalblaze" with bands of iron ore, ... 11 ft.

A small hole was bored from this point to the depth of 12 feet in the last named material, making the total depth attained 83 feet.

The work seems to have been directed by "Prof. Challis." The shaft is now filled with salt water.

Coal crops out at numerous places in the vicinity. It is said sometimes to show a thickness of two or three feet at the outcrop, but soon thins out.

Mr. Patton, on the east side of the river, near the south line of section 22, has made an excavation for coal and found a seam 18 inches thick which is tolerably hard.

The sandstone taken from the quarry above Flushing, is a pale, bluish rock, abounding in scales of white mica, ferruginous streaks, pyrites, carbonaceous streaks and curls, and much oblique lamination. What is quite remarkable, I saw in a block of this stone, in the vault of the Bank in Flint, a long club of fibrous talcose slate, a mineral said to occur in considerable abundance. This rock does not answer to the characters of the Woodville sandstone at any point where its identity is undoubted, and I am induced to regard it as a sandstone included in the coal measures. If it is so, this is the only instance within my knowledge where any of the included sandstones have attained sufficient development to be worked. It is likely, however, that the gray, homogenous, fine, gritty, faintly-banded sandstone, found within a mile or two of the city of Lansing, will be found to hold the same position.

Sandstone, not unlikely the Woodville sandstone, is found outcropping in the township of Montrose, on the borders of Saginaw county.

The next observed point in the boundary line of the coal field is near the village of Tuscola, in Tuscola county. On the S. W. ¼ Sec. 29, T. 11 N. 9 E., a seam of coal crops out in the bank of the Cass river. Numerous fragments of an arenaceous fire-clay, filled with *Stigmaria* roots, are strewn about. Some shales occur here, in which is found a *Lingula*.

According to information recently received from Dr. G. A. Lathrop, to whom I am under great obligations for his free co-operation in my researches, a shaft has been sunk on the north side of the river, with the following results:

Clay,	14 ft.
Fire-clay,	3 ft.

Dark shale,	4 in.
Coal similar to cannel,	8 in.
Bituminous coal,	2 ft.
Fire-clay,	4 ft.
Shale continuing at bottom,	8 ft.

One-half mile east of this, on the south side of the river, a hole was bored with the result as follows:

Sandstone,	18 ft.
Coal,	$4\frac{1}{2}$ ft.
Shale,	11 ft.
Sandstone continuing,	1 ft.

The outcrop of these strata traced northward, would probably strike the bay shore in the vicinity of Sebewaing.

From Barry, in Jackson county, around the northwest border of the basin, the boundary is still less perfectly known. The coal strata are known to outcrop, however, on Sec. 22, in the township of Benton, Eaton county, near the mouth of Grindstone Creek, and still again near the mouth of Coal Creek, in the same county. Indications also exist of the neighborhood of an outcrop near the center of Ionia county; but beyond this no authentic observations have been made. Coal is reported, however, to occur near the Big Rapids of the Muskegon, in Mecosta county. As a white quartzose sand, suitable for glass, is stated to occur at the Big Rapids, there is no improbability of the approach of the coal basin to that vicinity, for there are no such sandstones known except those which immediately overlie and underlie the coal series.

Numerous outcrops are known within the area marked out above. In the township of Lock, Ingham county, the coal has been taken from the bank of the Red Cedar river. This point is nearly in the straight line passing through Tuscola, Flushing and Jackson, and possibly like these points, occupies a position on the confines of the basin. As the strike of the underlying rocks, however, bends considerably toward the east, it is not unlikely that the eastern boundary of the coal basin will be found passing through Livingston county.

In the vicinity of Corunna, which is still further within the basin, the coal has been successfully worked on a small scale for a number of years. From an outcrop in the bank of a small creek on the W. ½ N. E. frl. ¼, sec. 22, Caledonia, Mr. Alexander McArthur has removed several thousand bushels of broken coal for the supply of neighboring blacksmiths. An excavation made at this place by the Detroit and Milwaukee Railroad Company, reveals the following section:

Highly ferruginous loam,	4 ft.
Blue clay, intersected by undulating bands of kidney iron nodules,	4 ft.
Black bituminous shale,	2 ft.
Bituminous coal,	3¼ ft.
Clay.	
Sandstone.	

The coal is of a handsome quality, and is intersected, like the seam in Jackson county, by a streak of pyrites.

Half a mile east of here, the overlying Woodville sandstone makes its appearance. In a shaft sunk by Frazer and Stanton, not the one now worked, the following section was passed:

Superficial materials,	5 ft.
Sandstone,	5 ft.
Clay,	5 in.
Coal,	3 ft.
Fire-clay, intersected by a band of impure, argillaceous iron ore,	16 ft.
Sandstone,	3 in.
Fire-clay,	4 ft.

The prevailing sandstone exposed at Rock Bar, at Blossom's quarry, and numerous other localities in the vicinity of Corunna, is probably the sandstone which overlies the coal.

At Owosso, a shaft was sunk by the Detroit and Milwaukee Railroad Company, with the following result, as communicated by B. O. Williams, Esq., to whom I am indebted for much assistance:

Sandstone, flesh colored, seen also in bed of river,	14 ft.
Black band iron ore,	1 ft.
Cannel coal,	2 ft.
Dark shales wth vegetable remains,	15 ft.

Black band,.. 8 in.
Bituminous coal,.................................. 3 ft.
Bluish soft clay,.................................. 1 ft.
Shales and arenaceous fire-clays, (as ascertained by boring,) the lower part black, terminating in a hard pyritiferous stratum,............................ 148½ ft.

Near the mouth of Six Mile Creek, in the township of New Haven, numerous explorations have been made. A shaft sunk at the mouth of the creek, furnished the following section, according to information received from Mr. George Ott. The shaft was sunk by Messrs. Silliman and Walker, 27½ feet, and the section continued by boring:

Superficial materials,.................................. 11 ft.
Clay, bluish black,.................................. 9 ft.
Calcareo-argillaceous black band,.................... 2 ft.
Cannel coal,.................................. 2 ft.
Clay and *coal,*.................................. 3½ ft.
Coal,.................................. 5 ft.
Clay, light colored,.................................. 2½ ft.
Coal,.................................. 2 ft.
Clay, light colored,.................................. 1½ ft.
Coal,.................................. 2 ft.
Clay, light colored.

A sandstone is seen in the vicinity, overlying the bluish black shale. The black band outcrops in the bed and bank of the Shiawassee river at the bridge, and has been quarried for building purposes. Half a mile up the Creek it is seen presenting a compact, fine-grained, calcareous character, of very black color, and seems capable of taking a polish. According to the statements given above, we have here 11 feet of coal within a vertical thickness of 18½ feet. The shafts which I have visited have generally been found filled with water, so that it has been impossible for me to make personal observations. In such cases I have deemed the statements of persons who watched the progress of the work as better than an entire absence of information. At this place, as in most others, I had the opportunity to inspect samples taken out. As to the nature of the strata, therefore, I have judged for myself, while for their thickness, I have had to depend upon others. If the information obtained from Mr. Ott is correct, (and it is corroborated by Mr. B. O Williams,) Six Mile Creek furnishes the greatest thickness of workable coal that has yet become known in the State.

Next to this, the greatest known thickness of any single vein of coal is 4 feet 1 inch, in a shaft and boring sunk on Sec. 35, in the township of Delta, Eaton county. My only knowledge of this locality is recently obtained from Messrs. J. A. Kerr and LaRue, of Lansing, who furnished me from their records the following statement of rocks passed through:

Superficial materials,	5 ft.	
Fire-clay, soft,	2 ft.	8 in.
Coal,	2 ft.	3 in.
Clay, somewhat bituminous,	4 ft.	3 in.
Coal,	1 ft.	11 in.
Fire-clay, white and hard,	5 ft.	8 in.
Argillaceous shale,	16 ft.	2 in.
Coal,		8 in.
Argillaceous shale, with some pyrites,	12 ft.	
Sandrock,	4 ft.	2 in.
Coal,	4 ft.	1 in.
Sandstone, grayish, soft.		

We have here a total thickness of 8 ft. 11 in. of coal, distributed in bands, as follows:

Coal,	2 ft.	3 in.
Clay,	4 ft.	3 in.
Coal,	1 ft.	11 in.
Argillaceous strata,	21 ft.	10 in.
Coal,		8 in.
Intervening strata,	16 ft.	2 in.
Coal,	4 ft.	1 in.
Total,	51 ft.	2 in.

The coal also outcrops at Chesaning, on the land of Sheriff Turner, and at several other points along the river in this township and St. Charles.

In the salt borings on the Saginaw river, coal is struck between 120 and 140 feet. At East Saginaw, according to notes of Dr. Lathrop, the following was found to be the section through the coal measures:

Alluvial and drift materials,.................... 92 ft.
Brown sandstone, (Woodville,)..................... 79 ft.
Shales, dark colored above, light below,............... 40 ft.
Bituminous coal,..................................3 or 4 ft.
Highly arenaceous fire-clay and sandstones,........... 20 ft.
Shales, below, dark, bituminous,................. 12 ft.
Sandstone, with thin seams of coal,.................. 10 ft.
Shale, ... 38 ft.
White sandstone, (Parma,)..........................105 ft.

The whole thickness of the coal measures here, between the overlying and underlying sandstones, is thus shown to be 123 feet, which is the greatest thickness yet measured. Probably, however, the thickness is still greater at Owosso.

Putting all the observations together, (of which it is not necessary to make any further details,) it appears that the rocks of the coal measures occupy a shallow basin, the longest axis of which is nearly coincident with the axis of Saginaw bay. This bay breaks over the northeastern rim of the basin, and near its head the rocks will probably be found to exhibit their greatest depression. It is not likely, however, that this depression varies greatly between Saginaw and Ionia county. In other words, the lowest depression of the carboniferous trough lies beneath a line extending from Ionia county into Saginaw bay. Along this line the coal measures will be found to have the greatest thickness, and the coal seams will be developed in greatest number and force.

When we speak of the carboniferous basin or trough, it must nevertheless be remembered that all these rocks repose very nearly in horizontal planes, so that the slight undulations into which they have been thrown by gentle disturbances since their solidification, have presented eminences and ridges which have subsequently been more or less worn down. It follows, therefore, that the Woodville sandstone is not everywhere found covering the coal measures, even within the area that has been described. The denudation has sometimes extended entirely through this sandstone, or into the shales below, or even so far as to reach below the coal seams. The tracts, however, within the carboniferous area, which have been entirely denuded of coal, must be very limited, so that in general terms, the whole area will be found productive.

From the numerous sections which have been given, it appears that one persistent seam of coal runs through the whole formation. This ranges in thickness between three and five feet, being thinnest near the borders of the basin. Toward the central axis of the basin, all the members of the series thicken, and several accessory seams of coal make their appearance. When this occurs, one of the seams is a cannel coal about 2 ft. in thickness. Immediately above this seam is a belt of black band, becoming in places highly calcareous, and passing into a black ornamental limestone or marble. To present the general structure of our coal measures more clearly to the mind, we may make use of the following table:

E. Bituminous shales and light clays,..................40 ft.
D. Black band passing into black limestone,............ 2 ft.
C. Bituminous and Cannel Coal in one or more seams, with aggregate thickness of 3 to 11 feet,..............11 ft.
B. Fire-clays and sandstones,..........................23 ft.
A. Shale, clay, sandstone and thin seams of coal,.......50 ft.

The shales of the coal measures are well stocked with the remains of the terrestrial vegetation. Fern leaves, in a beautiful state of preservation, are sometimes found in the black band. But few marine fossils occur, and these have been already noted.

17.—*Woodville Sandstone.*

Some account of this formation has necessarily been embraced in the description already given of the Coal Measures, and I shall add but few observations. Wherever it is not denuded, it is the capping stone of the coal measures. It is a friable, rather coarse, quartzose sandstone, stained to a variable extent with oxyd of iron. At Jackson, the rock is nearly white, and has been used in the manufacture of fine glass; at Corunna it is pale buff, and embraces abundant rich nodules of kidney iron ore, which, on the disintegration of the rock, are left in the soil; at Owosso it is flesh colored; near Lyons, in Ionia county, it is striped and mottled with red, or even of a uniformly brick red color. It varies equally in hardness, being sometimes sufficiently solid for grindstones and building stones. The State Prison is built of a rock supposed to be this. The material for the county offices at Ionia was also derived from the same source.

This rock embraces numerous comminuted remains of vegetation, and some well preserved stems of *Calamites* and *Lepidodendron.*

Although I have treated separately of the Parma Sandstone, the Coal Measures, and the Woodville Sandstone, there is no doubt that they all belong strictly to one geological epoch, and constitute what, in a more extended sense, may be designated the coal measures. This remark, however, is somewhat more applicable to the Woodville than the Parma Sandstone.

18.—*Superficial Materials.*

No traces have yet been discovered in the Lower Peninsula, of any of the geological formations intervening between the Coal Measures and the Boulder Drift. Drift materials are strewn over nearly the entire surface, and constitute a very serious obstacle in the way of the investigation of our geology. A large number of facts and observations is on hand as data for the discussion of this formation, but it will be necessary to content ourselves with a few general remarks.

Numerous evidences exist of the movement of heavy bodies over the underlying rocks, previously to their burial by the Drift. Wherever considerable surfaces are found exposed, they are seen smoothed and striated in the manner usually attributed to drift agency. The most remarkable examples are seen upon the Helderberg limestone at Brest, Stony Pt., and Pt. aux Peaux At Stony Pt., the surface of the limestone has been denuded of soil by the action of the waves, over an area of several acres. The whole surface is level, smooth and floor-like, and covered with a set of striæ running in perfectly parallel lines N. 60° W. One deep groove is seen belonging to this set The most remarkable feature seen here, however, is the occurrence of two parallel grooves crossing the first set and bearing N. 60° E. These grooves are 4 ft. $6\frac{1}{2}$ inches apart, $1\frac{3}{8}$ inches deep, 2 inches wide, and 25 feet long, issuing from under the cover of diluvial materials, and terminating at the point to which the waves have broken away the rock. The first impression which irresistibly forces itself upon the mind, is the conviction that a loaded wagon has been driven over the surface while in a yielding condition; and a couple of grooves parallel to these, seen for a part of the distance like the tracks of the second pair of wheels, greatly confirms the illusion.

The Island of Mackinac shows the most indubitable evidences of the former prevalence of the water, to the height of 250 feet above the present level of the lake; and there has been an unbroken continuance of the same kind of aqueous action from that time during the gradual subsidence of the waters to their present condition. No break can be detected in the evidences of this action from the present water-line upward for 30, 50 or 100 feet, and even up to the level of the grottoes excavated in the brecciated materials of "Sugar Loaf," the level of "Skull Cave" and the "Devil's Kitchen."

While we state the fact, however, of the continuity of the action during all this period, it is not intended to allege that the water of the lakes, as such, has ever stood at the level of the summit of Sugar Loaf. Nor do we speak upon the question whether these changes have been caused by the subsidence of the lakes, or the uplift of the island and adjacent promontories. It is true that the facts presented bear upon these and other interesting questions, but we must forego any discussion of them.*

*Abundant evidences are furnished along the shores of Lake Huron, of the unbroken continuity of the action of those physical forces which have transported and assorted the materials of the Drift. From the shingle beach formed by the violence of the last gale, we trace a series of beaches and terraces, gradually rising as we recede from the shore, and becoming more and more covered with the linchens and mould and forest growths which denote antiquity, until, in some cases, the phenomena of shore action blend with the features which characterize the glacial drift. These observations tally so well with the views of Pictet on the continuity of the Diluvian and Modern Epochs, as established by palæontological evidences, that I cannot forbear referring the reader to an article of his which falls under my notice as this report is going through the press. See *Bibliotheque Universelle de Geneve*, *Vol. VIII.*, *p.* 255. *Also*, *Silliman's Journal*, [2] *XXXI*, 345.

Upon the smooth and striated surfaces of the rocks, has been brought an immense deposite of waterworn and comminuted materials, derived from the breaking up and disintegration of pre-existing strata. We generally—almost universally—find the face of the rock overspread with a confused mixture of blue clay and azoic and plutonic boulders and pebbles. These coarse materials are often arranged in rude courses which have a curved or irregular dip, and may often be seen outcropping on a hill-side, or even upon the plain. At East Saginaw these materials are 90 to 100 feet below the surface. At Detroit they lie 130 feet below the surface. Through the interior of the State they are found outcropping at irregular intervals, producing occasional patches of ground principally noteworthy for their cobble stones. A field was noticed in the southern part of Jackson county in which, by measurement, the average distance between adjacent stones was only four inches. This small field had already furnished many hundred cords of these stones; but every plowing seemed to favor the development of a new crop. Strange to say, this and similar lands are found to produce excellent crops of wheat.

Great use is made of these cobble stones for purposes of paving in the cities, a use for which their great hardness and toughness renders them eminently fit. Mineralogically, they consist mostly of rounded fragments of syenite, greenstone, vitreous and jaspery sandstones, and hornblendic, talcose, and serpentinous rocks of the azoic series.

Above the boulder bed we find a deposite of argillaceous and arenaceous materials more distinctly stratified and assorted, as if by the action of eddying waters. So far as I have observed, the lake ridges and terraces are worked in these materials. Here we find buried, numerous tree trunks, generally of the White Cedar, many of which may be seen projecting from the bank which overhangs Lake Huron, near Fort Gratiot, and at numerous other points on the lakes.

The materials of this assorted drift are not so exclusively of extreme northern origin as those of the boulder drift. Perhaps two-thirds of the whole has been derived from the destruction of rocks within the Lower Peninsula; while a large and characteristic portion comes from the strata in the immediate neighborhood. The vicinity of a coal outcrop has filled the subsoil with fragments of coal, which can be traced, gradually diminishing in abundance, for one, two or ten miles. The experienced observer, however, is able to tell whether the source of the materials is near or remote, for the further they have been transported the more uniformly they become scattered amongst the other materials, while in the immediate vicinity of the outcrop the carbonaceous debris is not only *more abundant* but contains *more fine matter*, and is disposed *in streaks*. In a similar manner the vicinity of a limestone formation produces a calcareous soil; sandstone an arenaceous one; shale an argillaceous one. Nowhere is the connection between the soils and geological structure better shown than in Michigan. Even the arboreal vegetation of the peninsula is distributed in belts across the State, corresponding to the calcareous, arenaceous and argillaceous belts of soil overlying the corresponding rocks.

To this epoch of the drift seems to belong a bed of lignite discovered on Grand Traverse Bay, near the outlet, on the north shore. The following section embraces the lignite and associated beds:

F. Very fine yellow sand,............................12 ft.
E. Small boulders, pebbles and coarse sand with shells of *Melania*, and *Physa*,.................................... 7 ft.
D. Arenaceous clay, bituminous, soft, and somewhat plastic,.. 2 ft.
C. Lignite, dark brown, containing woody stems, (white cedar?) becoming below, a highly bituminous clay,.. 3 ft.
B. Clay, dark gray, very tough, with a few grains of sand and small pebbles, and considerable bituminous matter,..2½ ft.
A. Comminuted green shale, passing above into green clay, 2 ft.
Still lower, but not seen in this section, is a bed of green shale, lying above the black bituminous shale,......13 ft.

The lignite is compact, bituminous, and highly combustible.

The bed is traceable along the shore for a quarter of a mile, but nothing more is known of its extent. A single fragment was seen on the opposite shore of the bay.

The bed E, above the lignite, has evidently been deposited by the action of fresh water, since the epoch of the lignite; while the bed B, beneath the lignite, may belong to the same or may represent the glacial or boulder epoch.

Several other lignite beds are known upon the shore of Lake Superior, and I am informed that the inhabitants of that region are beginning to learn their use.

The inequalities left in the surface of the assorted drift, upon the withdrawal of the submerging ocean, remained filled with water, which, by constant drainage to the sea, in connection with accessions of fresh water only, have become our numerous inland lakes. These for many ages have been gradually filling up from several sources. 1st. Rains have transported the finer materials from the surrounding hills into these little basins. 2d. Spring-waters, charged with calcareous matter, have not only supplied the lakes with that material, but have precipitated large amounts upon the bottom. 3d. Mingled with these calcareous sediments, the dead shells of fresh water molluscs have accumulated in very great abundance. The union of these calcareous materials has formed a deposite of marl, continually thickening.

Around the shallow margins of these lakes is always a belt, abounding in various forms of aquatic vegetation, which, decaying, form a deposite of vegetable matters, resting upon the marl, from the water's edge to the inner limit of vegetable growth. With the filling of the interior, the shallow belt extends toward the center, and the vegetable deposite continually encroaches upon the lacustrine area, until the whole lake becomes a peaty marsh, with a bed of marl at bottom. Subsequent accessions of vegetable and calcareous matters fill the interstices of the porous soil, exclude the standing water, and convert the reeking marsh into dry and arable land. We behold, at the present day, these changes in all stages of progress.

The beaver and the muskrat may exert some agency in the inundation and drainage of lands, but a few observations upon the borders of our lakes will suffice to show that they are by no means the principal agents.

The beds of marl and peat thus accumulated constitute almost exhaustless repositories of nutritive matter for the recuperation of the hill-side soils, that have been exhausted of their soluble ingredients by the leaching rains, and an improvident system of farming. A consideration of the manner of preparing and applying these materials would be exceedingly interesting, but must be postponed for a final report.

Imbedded in these accumulations of marl and peat, are found the remains of the *Elephant*, *Mastodon* and *Elk*, the two former of which are now extinct from the continent, and the latter is only seen rarely in the remoter portions of the State. A fragment of a molar of the Mastodon was found by Dr. Miles at Green Oak, in Livingston county. A perfect molar of an elephant has been exhumed in ditching in the northern part of Jackson county. Other remains occur in Macomb county. By far the most interesting discovery has been made by Mr. G. M. Shattuck, in the township of Plymouth, in Wayne county. Mr. Shattuck here exhumed nearly an entire set of teeth of a Mastodon, including a piece of one of the tusks several feet in length. Some of these remains were in too friable a condition to be preserved, and others were injured by the injudicious handling of visitors. I have only had the opportunity as yet, of seeing five teeth. These prove to be the molar teeth from the lower jaw of *Mastodon giganteus*, three being from the left side and two from the right. The anterior one from the left side, is the single permanent premolar, and the posterior two, like the two on the right side, are the first and second true molars—the third, which is the largest of all, not appearing to have been developed at the time of the animal's death. These teeth are all in a beautiful state of preservation, still retaining their glossy enamel, and most of the fangs which belong to molar teeth. The tubercles of the crowns of the teeth were but little worn, showing, together with the absence of the third or largest true molar, that the animal had scarcely attained full maturity. The dimensions of the teeth were not extraordinary for proboscideans, being from two inches to four or five inches in antero-posterior diameter along the crowns, while the third molar of an adult Mastodon ranges from 7½ to 8½ inches in the same dimension. These interesting relics of a former age and a former population, are retained in the hands of their discoverer. It is greatly to be hoped that he will not allow them to become scattered or destroyed.

During the progress of the former survey, a large vertebra was discovered in the western part of the State which was recognized at the time as the caudal vertebra of a whale, by Prof. Sager, then State Zoologist.

CHAPTER IV.

General Observations—Table of Geological Formations

Many interesting considerations present themselves on a general review of the geology of the peninsula. From the Lake Superior Sandstone to the close of the Helderberg period, our State seems to have had a common history with Canada West, and the States on both sides of us. The same groups of rocks are traced uninterruptedly from New York across the peninsula of Canada to Michigan, and even to the Mississippi river, preserving throughout that whole extent as great a degree of palæontological identity as could be expected of faunas stretching over so many degrees of the earth's surface. It is true, as has been long since shown by Prof. Hall, that nearly every member of the Silurian and lower Devonian systems, thins gradually in its westward prolongation, loses somewhat of its arenaceous or argillaceous character, and becomes at the west much more calcareous—changes which have generally been regarded as proving the origin of the materials of those groups to have been at the east. It is interesting to observe, however, notwithstanding this westward attenuation, how completely we are able to recognize all the essential features of the New York System in our own State.

From the close of the Helderberg period, on the contrary, Michigan has had a history to some extent peculiar. The rocks of the Hamilton group can indeed be traced almost continuously from New York into our own State, but the palæontological characters are found materially changed, and the strata are more argillaceous. The Portage Group, of New York, supposing it to be represented by our Huron group, has received great accessions of argillaceous matter, and seems to have been deposited under circumstances more unfavorable to the existence of animal life The Chemung Group, supposed to be represented by our Marshall Group, has been traced uninterruptedly into Ohio, where it becomes almost non-fossiliferous. The Marshall Group is totally isolated from rocks of the same age anywhere beyond the limits of our peninsula; and though the sandstones bear some physical resemblance to those of the Chemung Group, of Ohio and New York, our formation contains little or no argillaceous matter; its fauna is remarkably rich, and its species are nearly all peculiar. The Napoleon Group, if correctly separated from the Marshall Group, has no distinct equivalent in surrounding States; and its entire destitution of organic remains will cause its true geological relations to remain in doubt.

If anything were wanting to show that the geological column in Michigan has been built up as a distinct and independent structure, the existence of the gypseous or Michigan Salt Group, supplies the deficiency. But even further than this, no obvious parallelism has yet been traced between the overlying carboniferous limestone, and the groups of this system further west. The indications already pointed out, however, lead to the conjecture that our limestone was accumulating during several of the epochs into which geologists have divided this period, though the isolation of our sea has resulted in little correspondence of organic remains. The paucity of rock-producing materials seems to have continued through the epoch of the coal—our measures not attaining one-twentieth the thickness of the same rocks in Ohio. The evidences lead us to the conviction that the Ohio and Michigan coal basins were never continuous, and that the waters did not flow over the separating ridge between the close of the Helderberg period and the Drift. It cannot be denied, however, that, supposing the carboniferous sea to have been a general one, the remoteness and comparative isolation of the Michigan bay, furnished occasion for great contrasts in stratigraphical, lithological and palæontological characters.

One other class of facts must be referred to, which weigh in the same direction. They constitute evidences that the materials for our upper Devonian and carboniferous rocks have been derived from the north. The Helderberg limestones are 350 feet thick at Mackinac, and not more than 60 feet thick in Monroe county. The Hamilton Group, so well developed in Thunder and Little Traverse Bays, is not recognized in the southern part of the State. The Huron Group with its gritstones and flagstones at Pt. aux Barques, contains only two strata of flagstone at Grand Rapids. The conglomerate at the base of the Marshall Group, at Pt. aux Barques, is recognized at none of the southern outcrops. The pebbles scattered through the Marshall and Napoleon Groups in Huron county, are entirely wanting in Jackson and Calhoun counties; while, on the contrary, extensive patches of the Marshall sandstone are found finely cemented by calcareous matter at Battle Creek, Jonesville and other southern points.

If our later palæozoic rocks are entirely isolated from those of adjoining regions; if their lithological characters are different; if their organic contents are peculiar; if their materials have been received from another direction; what prevents us from saying that Michigan has had a little geological history of her own, that her boundaries were marked out many thousand years ago—in short, that she was the very first of the States to take her place in this great and imperishable Union.

One other remark is suggested by this review of our rocks. The geology of Michigan discloses little connection between the Carboniferous Limestone and the Coal Measures; while the transition to Devonian rocks is imperceptible. I see no reason for drawing the broad lines which separate great systems, between the Marshall and Napoleon groups, or between the Napoleon group and the Carboniferous limestone. On the contrary, I see this limestone characterized by a peculiar, persistent, marine fauna, while the Parma Sandstone, the Coal Measures and the Woodville Sandstone, were accumulated in shallow waters near shores, or even in marshes; and are characterized, from bottom to top, by evidences of the proximity and abundance of terrestrial vegetation. These contrasts hold throughout the country, and in all countries. Whatever marine remains are found in the coal measures, belong to species distinct from those in the Carboniferous Limestone; and if the generic distinctions are not complete, the organic facies of one is vegetable and terrestrial; that of the other, animal and marine. Downward the types of the lower Carboniferous rocks descend into the upper Devonian—some carboniferous species, and numerous carboniferous types, even reaching the Hamilton group. Observations in Michigan suggest rather to draw the broad systematic lines below the Hamilton group, and between the Carboniferous Limestone and the Coal Measures.

SYOPTICAL VIEW OF THE GEOLOGY OF THE LOWER PENINSULA OF MICHIGAN.

V.—QUATERNARY SYSTEM.

(*c*) Soil—Peat, Marl, Calcareous Tufa, Bog Iron ore, Ochre Beds.

(*b*) Lake and river terraces, and other phenomena of altered drift; Lignite beds of lakes Michigan and Superior; Buried tree trunks.

(*a*) Boulder Drift; Diluvial striæ.

IV.—CARBONIFEROUS SYSTEM.

16. Woodville Sandstone, 79 feet; Jackson, Woodville, Barry Shiawassee county; Lyons; Tuscola county, &c.

15. Coal Measures, 123 ft.; consisting of

(*e*) Bituminous shales and clays, 40 ft.

(*d*) Black band, passing into black limestone, 2 ft.

(*c*) Bituminous and Cannel coal in one or more seams, with aggregate thickness of 3 to 11 ft.

(*b*) Fire clay and Sandstone, 23 ft.

(*a*) Shale, Clay, Sandstone and thin seams of coal, 50 ft.

14. Parma Sandstone, 105 ft.; Jackson county and salt borings at Saginaw.

13. Carboniferous Limestone, 66 feet:

(*c*) Upper, 10 ft.; Grand Rapids, Bellevue, Parma, Spring Arbor, Wild Fowl Bay, Charity Islands, Pt. au Gres.

(*b*) Middle, or Red Layer, 5 feet; Grand Rapids, Bellevue, Sandstone, Spring Arbor.

(*a*) Lower, 51 feet: seen at most of the above localities. Becomes arenaceous below.

12. Michigan Salt Group, 184 feet:

(*c*) Carbonaceous and argillaceous shale, gypseous and pyritous marls.

(*b*) Shales, marl, magnesian and silicious limestone, and thick beds of gypsum. The shales impregnated with salt.

(*a*) Saliferous shales and alternating arenaceous limestones.

11. Napoleon Group, 123 feet:

(*d*) Shaly micaceous sandstone, 15 feet.

(*c*) Napoleon sandstone, 78 feet, highly saliferous in many localities; Napoleon, Grandville, Rush lake, Pt. aux Barques.

(*b*) Shaly micaceous sandstone, 15 feet: Salt borings.

(*a*) Clay or shale, 15 feet, [more than 64 feet at East Saginaw (?)]

III.—DEVONIAN SYSTEM.

10. Marshall Group, (Chemung,) 159 feet:

(*c*) Reddish, yellowish and greenish sandstones, 147 feet: Marshall, Jonesville, Hillsdale, Battle Creek, Holland, Pt. aux Barques.

(*b*) Shaly micaceous sandstone, 10 feet: Jonesville, &c.

(*a*) Conglomerate, 2 ft.: Grindstone Quarries, Pt. aux Barques.

9. Huron Group, (Portage), 224 feet:

(*d*) Fine bluish gritstones, 14 feet: Pt. aux Barques.

(*c*) Shales, limestones and flagstones, 18 feet. The Kidney Iron clays of Branch county are supposed to belong here. Shore of Lake Huron, below Pt aux Barques; Branch, Calhoun, Kalamazoo and Lenawee counties.

(*b*) Green shale, 10 feet: Grand Traverse Bay.

(*a*) Black bituminous shale, 20 feet: Sulphur Island, Squaw Pt, Grand Traverse Bay.

8. Hamilton Group, 55 feet:

(*c*) Crystalline limestone, with included lenticular clayey masses, 23 feet: Partridge Pt., Little Traverse Bay.

(*b*) Argillaceous limestones, eminently fossiliferous, with alternating shales, 17 ft.: Partridge Pt., Little Traverse Bay.

(*a*) Black bituminous limestone, 15 feet: Carter's quarry, near Alpena; Thunder Bay Island, Little Traverse Bay.

7. Upper Helderberg Group, 354 feet:

(*e*) Brown, bituminous limestone, 75 feet: Monroe, Presque Isle and Emmet counties.
(*d*) Arenaceous limestone, 4 feet: Monroe county, Crawford's quarry.
(*c*) Oolitic limestone, 25 feet: Bedford, Raisinville, &c., Monroe county; Mackinac.
(*b*) Brecciated limestone, 250 feet: Stony Pt., Pt. aux Peaux, Mackinac and vicinity.
(*a*) Conglomerate, cherty, and sometimes agatiferous, 3 feet: Mackinac, Sitting Rabbit.

II.—UPPER SILURIAN SYSTEM.

6. Onondaga Salt Group, 37 feet:

(*d*) Chocolate colored limestone, 10 feet: Monroe county, Mackinac.
(*c*) Calcareous clay, 3 feet: Bois Blanc.
(*b*) Fine, ash colored, argillaceous limestone, with acicular crystals, 14 feet: Monroe county, at Montgomery's quarry, Ida; Otter Creek and Plumb Creek quarries; Mackinac; Round and Bois Blanc islands.
(*a*) Variegated, gypseous marls, with imbedded masses of gypsum, 10 feet: St. Martin's islands; Little Pt. au Chene.

5. Niagara Group, 97 feet:

(*g*) Thin-bedded brown limestone, 6 feet: south side Drummond's Island.
(*f*) White, massive, crystalline limestone, 20 feet: south and southeast sides Drummond's Island; coast west of Detour.
(*e*) Rough, vesicular limestone, 6 feet: east end of Drummond's Island.
(*d*) Limestone, in thin broken layers, 8 feet: Ibid.
(*c*) Limestone, geodiferous, rough, crystalline, 45 feet: Ibid.
(*b*) Limestone, hard, gray, crystalline, 7 feet: Ibid.
(*a*) Arenaceous limestone, 5 feet: Ibid.

4. Clinton Group, 51 feet:

(*c*) Argillo-calcareous limestone, very light colored, and evenly bedded, 14 ft.: E. and W. ends of Drummond's Island.
(*b*) Argillo-calcareous limestone, dark, containing geodes and gashes, 3 feet: N. E. side Drummond's Island.
(*a*) Alternations of argillaceous, bituminous and calcareous limestones, 34 feet: Ibid.

I.—LOWER SILURIAN SYSTEM.

3. Hudson River Group, (observed,) 18 feet:

Argillaceous limestone, filled with fossils in the upper part, 15 feet: N. side Drummond's Island.

Bluish-gray subcrystalline limestone, (observed,) 3 feet.

2. Trenton Group, 32 feet:

(*e*) Dark blue, subcrystalline limestone, with 3 feet of dark-green areno-calcareous shale above, 7 feet; North side Drummond's and St. Joseph's Islands.

(*d*) Dull-gray limestone, hard, silicious, 2 feet. Ibid.

(*c*) Blue, argillaceous limestone, 9 feet: Ibid.

(*b*) Limestone, dark, bluish-gray, with partings of green shale, 12 feet: Ibid.

(*a*) Limestone, gray, silicious, resting on quartz, 2 feet: Sulphur Island, north of Drummond's.

1. Lake Superior Sandstone, (Potsdam), at the Sault, 18 feet.

Total *observed* thickness of the Palæozoic rocks, 1,725 feet. Actual thickness probably 2,500 feet.

CHAPTER V.

Tables of deep borings in the State, with an exhibition of their Geology.

References have frequently been made to borings that have been executed in our State, in search of salt, coal or other valuable products; and isolated facts, obtained by such borings, have, in many instances, been incorporated into the preceding chapters. In the present chapter, I present connected and complete statements of the kind of rocks passed through, in most of the deep borings of our State. It has not been thought best to present these records in all their details; I have, therefore, greatly condensed them, taking care, however, to mention every important change in the strata. The first column in all the tables shows the depth of the well at the upper part of the stratum named in the last column. The second column shows the thickness of the stratum. When the several strata which constitute a formation or group, are passed, a line is drawn across the second column, and the total thickness of the formation or group is entered opposite, in the third column. The table at the end is a summary of the whole.

I.—Artesian Well at Detroit.

["During the years 1829–30, the Hydraulic Company, with a view of supplying this city with spring water, commenced and completed, (although without gaining the object intended,) an artesian well, near that point where Wayne Street intersects Fort Street. This point is elevated 36 feet above the level of the surface of Detroit river. The work was conducted under the direction of A. E. Hathon. In the North-western Journal for April 21, 1830, an article was published from the pen of that gentleman, of which the following is an abstract of the strata and depth, in the words of the article to which allusion is made:"—*Dr. Houghton's Notes.*]

At Depth of	Interven-ing Thick-ness.	Thickness of Forma-tions.	DESCRIPTION OF ROCKS, &c.
Ft.	Ft.	Ft.	
0	10		" Common Alluvion."
10	118		" Plaster Clay." [" Marly."—*Houghton.*]
128	2		" Common beach sand, with coarse gravel."
		130	
130	120		" Compact limestone." [Probably Heldorberg and Onondaga salt group together.—*W.*]
250	2		" Gypsum and Salt."
252	8		" Compact Lime." [Probably Niagara limestone.—*W.*]
260			

II.—State Salt Well, Grand Rapids. Sec. 3, T. 6 N., 12 W. Condensed from the records kept under Dr. Houghton's directions. Bored in 1841–2.

At Depth of	Interven-ing Thick-ness.	Thickness of Forma-tions.	DESCRIPTION OF ROCKS, &c.
Ft.	Ft.	Ft.	
0	40		Alluvial, &c., 5–6 feet clay, thin sand and gravel.
		40	
40	7		" Clay;" Gypsum 6½ feet.
47	1		" Very hard rock, supposed to be hornstone."
48	13		" Clay" and " slate" alternately, with 1–3 in. " hard rock," several times recurring in the lower 4 feet.
		21	
61	109		" Sandrock," " hard." At 63 feet, a spring, water brackish, cavity 3 in.; sandrock continuing; softer, with numerous cavities; brine strengthening; rock harder at 104 ft.
170	9		" Mixture of clay and sand—quite hard."
179	5		" Clay slate."
		123	
184	101		Hard sandrock, 19 ft.; cavities, water very salt; " soft sandrock" at 204–244 ft.; very hard at 245–246; soft, 247–278. At 265 feet, brine overflowing profusely, and increasing to 284 feet.
285	2		" Blue clay."
287	20		" Common sandrock."
307	24		" Ash colored clay and sandrock," " about equal parts."
331	12		" Sandrock, quite hard."
		159	
343	130	130	" Clayrock." Water doubled at 361, and somewhat stronger. From 417–421 very soft like blue clay, then a few black gravel stones, then shale.
473			" Clayrock." Continuing.

III.—LYON'S SALT WELL, GRAND RAPIDS, near Bridge St. Bridge, commenced January, 1840, and finished Dec. 25th, 1842, condensed from records kept under the direction of Hon. Lucius Lyon.

At Depth of	Interven-ing Thick-ness.	Thickness of Forma-tions.	DESCRIPTION OF ROCKS, &c.
Ft.	Ft.	Ft.	
0	13		Limerock, lower 9 feet geodiferous.
14	6		"Yellow sandrock." [Probably lower arenaceous beds of the limestone.]
		19	
20	2		"Blue Clay."
22	5		"Coarse, reddish sandrock."
27	47		Argillaceous beds, interstratified with gypseous deposits.
74	7		"Very hard, sharp-gritted, bluish sandrock." At 76 feet, fresh spring.
81	19		"Clayrock." "First indications of salt."
100	79		Argillaceous beds, sometimes "sandy," [pyritous grains,] sometimes gypseous.
179	1		"Hard sandrock." [These layers are also called "waterlime."]
180	11		Clayrock.
		171	
191	109		Sandrock, varying from "dark" and "hard," to "white" and "soft," (199 feet); dark blue (216); coarse, loose and reddish (248). Cavity of 6 inches, and great spring of water at 264¼ feet.
300	9		"Clayrock, intermixed with fine particles of sand."
		118	
309	66		Sandrock, varying between "hard," "coarse" and "loose."
375	14		"Clay and sandrock of about equal parts."
389	34		"Sandrock, coarse, loose,—of about an ash color."
423	12		"Clay and sandrock of about equal parts."
435	11		"Coarse, loose sandrock;" water doubled, and salter.
		137	
446	18		"Clayrock."
465	2		Sandrock.
467	194	214	Clayrock. One foot of sandrock at 495 feet.
661			Clayrock continuing. Temp. of water in well 50½ Fah.

IV.—Scribner's Salt Well, near the Railroad Depot, Grand Rapids. Bored in 1859-60. Notes furnished by James Scribner, Esq.

At Depth of	Intervening Thickness.	Thickness of Formations.	DESCRIPTION OF ROCKS, &c.
Ft.	Ft.	Ft.	
0	51		The portion of the Carboniferous Limestone below the "Red Layer."
		51	
51	1		Shale, compact.
52	2		Hard blue limestone, called here "Waterlime."
54	79		Argillaceous strata with occasional beds of limerock, and coarse and fine grained layers of sandstone, from 1 to 5 feet in thickness.
133	66		Argillaceous rocks, somewhat harder. First indications of salt. The whole series saliferous, gypsiferous and pyritiferous. Occasional layers of sandstone and limestone.
199	5		Highly ferruginous and pyritous rock, exceedingly hard.
		153	
204	54		Sandrock, porous, with salt water.
258	57		Sandrock.
315	10		Clay.
		121	
325	55	55	Sandrock.
380			Sandrock continuing.

V.—Powers and Martin's Salt Well. Grand Rapids, half mile N. W. from Scribner's Well. Samples of the borings were furnished for examination by Mr. A. O. Currier.

At Depth of	Intervening Thickness.	Thickness of Formations.	DESCRIPTION OF ROCKS, &c.
Ft.	Ft.	Ft.	
0	16		Superficial materials.
		16	
16	3		Limestone, light grayish buff, fine grained, with small disseminated crystals of spar.
19	5		Red, arenaceous limestone, passing above and below into the gray limestone.
24	32		Limestone, gray above, then somewhat pyritiferous, with alternating shaly layers. Toward the bottom, becoming arenaceous, and in places cherty.
		40	
56	7		Shale, black and carbonaceous, with grains of pyrites; becoming more arenaceous below.
63	10		Clay, light colored, effervescing.
73	8		Clay, light, with nodules of pyrites, and some streaks of a white, pulverulent substance, which effervesces.
81	2		Shale, dark greenish, somewhat indurated.
83	9		Clay, unctuous, pyritous, arenaceous.
92	5		Shaly grit, dark, carbonaceous.
97	1		Fire-clay, with streaks of snowy gypsum.
98	18		Shale, pyritous, arenaceous, gypsiferous—sometimes greenish.
116	4		Shale, filled with grit.
120	7		Shale, with angular fragments of chert and streaks of gypsum. Lower portion highly gypseous, and then dark brown.
127	9		Shale, abounding in quartzose sand. Between 131-5 feet, highly gypseous.
136	2		Sandstone, dark, shaly, very fine.
138	18		Shale, with gypsum and sand. *First salt water*. Below this, alternating clay and shale, both abounding in grit.
		100	
156			Siliceous limestone, very hard. Salometer 100°, the supply being half gallon per minute. It is not thought that any brine was obtained below 138 feet.

VI.—BUTTERWORTH'S SALT WELL, Grand Rapids, at his Foundry. Notes furnished by R. E. Butterworth, Esq. Bored in 1860.

At Depth of	Interven-ing Thick-ness.	Thickness of Forma-tions.	DESCRIPTION OF ROCKS, &c.
Ft.	Ft.	Ft.	
0	24		Limestone, 12 ft.; soft sandrock, 5 ft.; limerock, 1 ft.; clay slate, 1 ft.; limerock, hard, 5 ft.
24	3		Calcareous sandrock, soft.
		27	
27	30		Clay, 7 ft.; shale, 8 ft., fresh water; limestone and clay alternately, 15 feet.
57	4		"Brown, hard limestone," filled with spar. First salt.
61	61		Varying argillaceous strata, all saliferous, with occasional beds of gypsum.
122	7		Very hard limerock, with gypsum.
129	48		Argillaceous strata, with much gypsum. Brine 91 gal. to bushel of salt. (Butterworth). At 146 ft. brine 5°, Beaume, temp. 50° Fah.
177	7		Limerock and gypsum.
		157	
184	77		Sandrock, bluish-gray. Water increasing.
261	13		Sandrock, gray. Water suddenly gushing up at the rate of 350 gallons per minute.
274	19		Sandrock, argillaceous.
293	10		Dividing shale. Sal. 20° to 26°.
		119	
303	128		Sandstone with shaly partings. Sal. 20° to 26°.
		128	
431	59		Alternating shales and flagstones.
490			Same continuing.

VII.—SALT WELL OF INDIAN MILL CREEK SALT CO. Grand Rapids, 25 rods N. of Powers and Martin's. Bored in 1860. Notes furnished by Ball, Clay & Co.

At Depth of	Intervening Thickness.	Thickness of Formations.	DESCRIPTION OF ROCKS, &c.
Ft.	Ft.	Ft.	
0	81	81	Gravel, sand, &c , with 4 inches clay at bottom. [This well seems to have struck a fissure in the limestone, or a place where the limestone had been entirely denuded.]
81	3		Gypsum, white.
84	8		Clay and shale.
92	4		Hard rock.
96	34		Clay, generally soft. At 128 feet, *first brine*.
130	7		Hard sandrock.
137	4		Clay.
141	9		Coarse gravel (!)
150	7		Fine gravel. [Possibly the gravel, so called, consisted of grains and nodules of pyrites disseminated through the clay, as in Powers & Martin's well.]
157	4		"Sandrock," extremely hard. [Supposed to be the bottom of Powers & Martin's well.]
161	42		Gypsum and Clay.
203	2		"Black sandrock."
205	9		"Hard sandrock." [These are probably the "Waterlime" layers.]
		133	
214	105		Sandrock, soft. Brine, flowing 10 gallons per minute, at 19° Sal. At 295 feet, Sal. 27°.
319	15		Clay and shale
		120	
334	29		Sandrock.
363	2		Black Iron-sand.
365	59		Sandrock.
424	10		Streaks of clay and sandrock.
		100	
434			Clay. Discharge of water, 135 gallons per minute.

VIII.—J. W. Windsor's Salt Well near Grand Rapids. Locality—fraction No. 1, Sec. 12, T. 7 N., 12 W. Notes furnished by Mr. Windsor's Superintendent of operations. Well bored in 1860.

At Depth of	Interven-ing Thick-ness.	Thickness of Forma-tions.	DESCRIPTION OF ROCKS, &c.
Ft.	Ft.	Ft.	
0	43		Superficial.
		43	
43	21	21	Limestone. 8 in. clay and gravel at 53 ft. Drill went down rapidly 15 inches, at 57 ft.
64	8		Dark shale, with blue below, underlain by 8 in. hard limestone.
72	4		Sandstone, very hard, yellow and gray.
76	12		Shale, gypsum and clay.
88	1		Sandstone.
89	10		Clay, shale and gypsum.
99	9		Greenish clay and shales, with black streaks.
108	24		Gypsum, alternating with shale of varying hardness, and occasionally greenish.
132	20		Blackish-blue shale.
152	4		Gypsum.
156	10		Black shale. *First brine at 164 feet.*
166	13		Gypseous clay, very salt, underlain by black, salt shale, alternating with gypsum.
179	4		Black, very hard rock.
183	57		Dark flinty beds, interlaminated with clay and gypsum. Shale below.
240	8		Very hard, pyritiferous rock, with gypseous clays.
		184	
248	79		Sandrock. Brine 16° at 259 ft.—20° at 278 ft.—17° at 319 ft.
327	22	101	Clay and sandrock, followed by clay, with some very hard streaks. Sal. 26°.
349	74		Sandrock, white. Sal. 31° at 391 ft. The overflow, 24°.
423	23		Argillaceous sandrock, fine. Brine remaining the same.
		97	
446			Same continuing. Discharge of water about 35 gallons per minute.

IX.—Deep Boring for Coal. S. W. ¼ N. W. ¼ Sec. 36, Sandstone, Jackson county. Notes made mostly from samples preserved by John Holcroft, Esq. Greatly condensed.

At Depth of	Intervening Thickness.	Thickness of Formations.	DESCRIPTION OF ROCKS, &c.
Ft.	Ft.	Ft.	
0	4		Superficial materials.
		4	
4	24		Sandstone, varying from nearly white to yellow and ochreous, mostly incoherent.
		24	
28	15		Limestone, siliceous, (3 feet,) followed by ochreous sandstone, argillaceous sandstone (3 feet), calciferous sandstone with greenish streaks (2 feet), cherty limestone (2 feet), all which might come under carboniferous limestone.
43	.15		Sandstone, light, becoming argillaceous, micaceous, with partings of shale.
58	2		Limestone, brownish, siliceous, with green blotches (4 in.); shale, bluish dark (8 in.); limestone arenaceous (9 in.); hornstone (1 in.)
		32	
60	31		Argillaceous strata, not effervescing, with thin bands of micaceous, argillaceous sandstone, ending with 9 feet of unctuous clay.
91	11		Sandstone, with a few pebbles (3 feet); followed by alternating shale and ironstone.
102	7		Chert (6 feet); shale, with pyrites (1 foot).
		49	
109	94		Sandstone, bluish-white, quartzose—a powerful water-course at 127 feet, and another at 156 feet. In the lower half, less uniform, by turns colored, argillaceous, effervescing, coarser.
203	20		Clay, coarsely arenaceous, then finer.
		114	
223	56		Arenaceous strata, with occasional thin argillaceous partings; powerful water-courses at 269 feet and 278 feet; ending in a very hard sandrock.
279			Sandrock continuing.

X.—Hibbard's Artesian Well, Jackson. Notes furnished by William Walker.

At Depth of	Intervening Thickness.	Thickness of Formations.	DESCRIPTION OF ROCKS, &c.
Ft.	Ft.	Ft.	
0	18		Sandstone, (with some overlying soil).
18	6		"Light colored slate." [Argillaceous micaceous sandstone?]
24	4		Sandstone, with kidney Iron.
		28	
28	16		Shale, black, bituminous.
44	3		Bituminous coal.
47	16		Shale, containing ironstone above, and becoming fire-clay at bottom.
		35	
63	20		Alternations of fine and "cherty" sandstone, with blue and light colored "slate," which may have been a fine argillaceous sandstone.
83	18		Sandstone, varying from coarse to fine.
101			Water.
101	5		Sandstone.
		43	
106			Sandstone continuing.

XI.—East Saginaw Salt Co.'s Wells. About $\frac{3}{4}$ mile north east from the center of town, on the river, nearly opposite Carrolton. Condensed from records kept by G. A. Lathrop, M. D. Bored in 1859–60.

At Depth of	Intervening Thickness.	Thickness of Formations.	DESCRIPTION OF ROCKS, &c.
Ft.	Ft.	Ft.	
0	92		Alluvial and Drift materials. Salometer 1°.
		92	
92	79		Brown sandstone, with angular grains. Temp. 47°; Salometer 2°.
		79	
171	40		Shales, first dark, then light.
211	23		Sandstone, [highly arenaceous Fire-clay ?] and 3 or 4 ft. of *Coal*.
234	12		Shales, below, dark, bituminous.
246	10		Sandstone with thin seams of *Coal*.
256	38		Shales. Temp. 50°, Sal. 14°. Discharge 80 gallons per minute.
		123	
294	105		White sandstone.
		105	
399	65		Limestone, embracing 6 beds of "sandstone," from 6 in. to 2 ft. thick, (the uppermost 5½ ft.), and terminating in an arenaceous limestone with shaly matter.
		65	
464	3		Shales.
467	20		Sandstone. Sal. 26°.
487	29		Shales.
516	43		Shales, with intercalated sandstones 6 in.—2 ft. thick. Sal. 44°–60°.
559	10		Fine blue sandstone. ["Waterlime"?] Sal. 64° at 568 ft.
569	15		Dark shales.
584	11		Fine blue sandstone, ["Waterlime"?] 3½ ft. shale at 590½ ft.
595	3		Grayish, coarser sandstone, with angular grains.
598	7		Dark shales.
605	15		Sandstone, hard, becoming micaceous—at 610 ft. calcareous.
620	7		Dark shales.
627	6		Limestone, hard, brown.
		169	
633	109		Sandstone
669			Bottom of first well.
742	42		Red shale.
784	1		Blue shale.
785	18		Red shale.
803	3		Blue shale.
		173	
806			Bottom of second well.

TABLE

SHOWING THE THICKNESS AND DEPTH OF FORMATIONS AT VARIOUS POINTS.

FORMATIONS.	State Well, Grand Rapids.		Lyon's Well, Gr'd Rapids.		Butterworth's, G. Rapids.		Scribner's, Grand Rapids.		Powers and Martin's G. Rapids.		Ball, Clay & Co., Grand Rapids.		Windsor's, Grand Rapids.		East Saginaw.	
	Depth.	Thickness.	Depth.	Thickness.	Depth.	Thickness.	Depth.	Thickness.	Depth.	Thickness.	Depth.	Thickness.	Depth.	Thickness.	Depth.	Thickness.
	Ft.	Ft.	Ft.	Ft.	Ft.	Ft.	Ft.	Ft.	Ft.	Ft.	Ft.	Ft.	Ft.	Ft.	Ft.	Ft.
Superficial,	0	40	......	......	......	......	......	......	0	16	0	81	0	43	0	92
Woodville Sandstone,	......	......	......	......	......	......	......	......	......	......	......	......	......	......	92	79
Coal Measures,	......	......	......	......	......	......	......	......	......	......	......	......	......	......	171	123
Parma Sandstone,	......	......	......	......	......	......	......	......	......	......	......	......	......	......	294	105
Carboniferous Limestone,	......	......	0	19	0	27	0	51	16	40	......	Wanting.	43	21	399	65
Michigan Salt Group,	40	21	19	171	27	157	51	153	56	100	81	133	64	184	464	169
Napoleon Group,*	61	107 14	190	109 9	184	109 10	204	111 10	......	......	214	105 15	248	91 10	633	109 64
Marshall Group,	184	159	308	137	303	128	325	55	......	......	334	100	349	97	......	......
Huron Group,	343	130	445	214	431	59	......	......	......	......	......	......	......	......	......	......
Hamilton Group,	......	......	......	......	......	......	......	......	......	......	......	......	......	......	......	......
Helderberg Group,	......	......	......	......	......	......	......	......	......	......	......	......	......	......	......	......
Onondaga Salt Group,	......	......	......	......	......	......	......	......	......	......	......	......	......	......	......	......
Niagara Group,	......	......	......	......	......	......	......	......	......	......	......	......	......	......	......	......
Totals,	......	471	......	659	......	490	......	380	......	156	......	434	......	446	......	806

* In giving the thickness of the Napoleon Group, the parting shale is distinguished from the sandstones.

The last number given for each locality does not show the total thickness of the corresponding formation, but only the depth to which it was penetrated before the boring stopped.

TABLE

SHOWING THE THICKNESS AND DEPTH OF FORMATIONS AT VARIOUS POINTS.

FORMATIONS.	Saginaw City.		Carrolton.		Portsmouth.		Bay City.		Tuscola.		Detroit.		Adrian.		Hibbard's Well, Jackson.		Deep boring, near Woodville.	
	Depth.	Thickness.	Depth.	Thickness.	Depth.	Thickness.	Depth.	Thickness.	Depth.	Thickness.	Depth.	Thickness.	Depth.	Thickness.	Depth.	Thickness.	Depth.	Thickness.
	Ft.	Ft.	Ft.	Ft.	Ft.	Ft.	Ft.	Ft.	Ft.	Ft.	Ft.	Ft.	Ft.	Ft.	Ft.	Ft.	Ft.	Ft.
Superficial,							0	86			0	130	0	58	0	4	0	4
Woodville Sandstone,									0	18					4	24		
Coal Measures,									18	16					28	35		
Parma Sandstone,															63	43	4	24
Carboniferous Limestone,																	28	22
Michigan Salt Group,																	60	49
Napoleon Group,*						88	545	85 / 12									109	94 / 20
Marshall Group,																	223	56
Huron Group,													58	31				
Hamilton Group,																		
Helderberg Group,																		
Onondaga Salt Group,											130	122						
Niagara Group,											252	8						
Totals,								630				260		89		106		269

* In giving the thickness of the Napoleon Group, the parting shale is distinguished from the sandstones.

The last number given for each locality does not show the total thickness of the corresponding formation, but only the depth to which it was penetrated before the boring stopped.

CHAPTER VI.

ECONOMICAL GEOLOGY.

It is undoubtedly contemplated, that in the presentation of a final report upon our geology, prominence shall be given to the economical materials furnished by the earth's crust, embracing an explanation of the principles concerned in searching for them, plain practical rules of procedure, the best methods of extracting, purifying and preparing them, and the uses to which they may be applied.

In the preceding descriptions of our formations, I have, in many cases, made allusion to the uses to which the various rocks and mineral products of the Lower Peninsula seem to be well adapted; and as a general statement, I do not deem it best to do anything more than this in the present report. In regard to the leading mineral interests of the Lower Peninsula, however, it may be expected that I should furnish, even in a report of progress, a greater amount of data for the practical guidance of those interested. For the purpose of presenting a view of the variety of our mineral resources, I subjoin the following table:

CLASSIFIED LIST of Products of the Economic Geology of the State of Michigan, and of subjects connected with their description.

I. Metallic Ores.

1. *Ores of Iron.*
   - *a.* Iron Pyrites.
   - *b.* Mispickel.
   - *c.* Magnetite.
   - *d.* Hæmatite.
     - Specular.
     - Micaceous.
     - Red.
     - Red Ocher.
     - Red Chalk.
     - Jaspery Clay Iron.
   - *f.* Spathic Iron Ore.
   - *g.* Manufactured Iron.
   - *h* Associated Minerals.
2. *Copper and its Ores.*
   - *a.* Native Copper.
   - *b.* Copper Pyrites.
   - *c.* Erubescite.
   - *d.* Gray Copper Ore.
   - *e.* Chrysocolla.
   - *f.* Copper in process of Manufacture.

Clay Iron Stone.
Lenticular Iron Ore.
e. Limonite.
Brown Hæmatite.
Yellow Ocher.
Yellow Clay Iron Stone.
Bog Iron Ore.
g. Associated Minerals.
3. *Silver and its associates.*
4. *Lead and its associates.*
5. *Other Metallic Ores.*
6. *Fluxes used in the reduction of Ores.*

II. Coal.

1. Bituminous Coals.
2. Cannel Coals.
3. Associates of Coal.
4. Cokes.
5. Gas.

III. Building Stones.

1. Syenite and Granite.
2. Sandstones.
3. Limestones.
4. Gypsum.
5. Marble.

IV. Materials for Cements.

1. Quicklime.
2. Waterlime.
3. Gypsum.

V. Materials for Ornamental Purposes.

1. Gypsum.
2. Marble.
3. Chrysocolla.
4. Agate, &c.

VI. Materials For Paints.

1. Ocher.
2. Manganese.
3. Ferruginous shales.

VII. Gypsum.

1. As a fertilizer.
2. As a cement.
3. For architectural purposes.
4. For ornamental purposes.

VIII. Salt.

1. Geological relations.
2. Brine.
3. Salt.
4. Sections of Borings.
5. Statistics and Calculations.

IX. Clays.

1. For Fire-bricks.
2. For common Bricks, Tiles, &c.
3. For Pottery.
4. For Pipes.

X. Sand and Gravel.

1. For Mortar.
2. For Glass.
3. For Moulding.
4. For Bricks and Walls.
5. Stationer's Sand.

XI. Gritstones.

1. Materials for Grindstones.
2. Materials for Whetstones.
3. Materials for Hones and Oilstones.

XII. Lithographic Stones.

XIII. Materials for Roads and Walks.

XIV. Soils.

XV. Materials for improving the Soil.

1. Gypsum.
2. Marl.
3. Peat.
4. Brine.
5. Sand.
6. Clay.

XVI. Wells and Springs.

1. Common Wells and Springs.
2. Artesian Wells.
3. Mineral Waters.

Most of the materials embraced in the above enumeration are of the very best quality; and when the union of capital and intelligence shall have brought our resources to such a degree of development as they admit, Michigan will be seen to stand among the leading States in point of mineral wealth.

COAL.

Many facts have already been stated which have a direct economical bearing upon the search for coal. A few suggestions may here be added:

1. The occurrence of fragments of coal in the soil, or in excavations for wells, does not prove the existence of a coal seam within many miles, as the outcropping edges of all the rocks have been broken up, and the fragments distributed toward the south.

2. In the examination of loose fragments, it may be remembered that the nearer we approach the outcrop of the solid seam, the more abundant the fragments become, especially the finer ones, while at the same time they are *less equally* distributed through the soil.

3. The occurrence of an extensive nest of fragments may result from the destruction of a former small outlier of the coal basin, and may be detached many miles from the principal seam.

4. When an outcrop is actually found, it will frequently be seen to dip *away* from the coal basin, as if bent down at the margin. The miner should not be misled by this peripheral dip.

5. Such seam will be found, generally, thinner than at points nearer the center of the basin.

6. The coal will be found much changed and deteriorated by the action of the elements. The quality will be found to be improved at increased distances from the surface.

7. The structure of our measures is such that it is useless to dig or bore anywhere to a greater depth than thirty feet below any seam of coal two feet thick. All the rest, if any, will be embraced within that distance.

8. It should be remembered that there are black shales *below* the coal as well as *above.*

9. It should also be remembered that the overlying (Woodville) sandstone is not easily distinguished from the underlying (Parma) sandstone, while these two sandstones are essentially distinct—sometimes 123 feet apart, and sometimes, on the borders of the coal basin, only 15 feet apart.

10. At any point favorably situated in other respects, lying a few miles within the circuit which has been traced out, productive coal seams may be confidently sought for.

11. The great practical difficulty in working them will be found in their situation below the general level of the surrounding surface, so that the shafts and drifts will contain water. By using good judgment, however, locations can be selected sufficiently high to obviate any serious annoyances from this source.

12. Care must be exercised against being misled by the black bituminous shales of the northern part of the peninsula. They burn freely, and closely resemble the coal shales; but they lie five hundred feet below any seam of coal.

The qualities of our coals have not yet been scientifically tested. It should be done. The following, the only chemical analysis in my possession, is said to have been procured in New York, by Mr. Hayden, of Jackson, upon a specimen of cannel coal, from the shaft of the Jackson City Co.:

Analysis of Cannel Coal from Jackson.

Carbon,	45
Volatile matter,	49
Ash,	2
Water,	2
Sulphur,	2
	100

Of the bituminous coals, several qualities may be easily distinguished by inspection. Some samples, too carelessly quarried, retain a considerable quantity of pyrites, which, on heating, gives off its sulphur, which becomes an annoyance in domestic use, and a positive detriment for mechanical purposes. Other samples, taken at points near the outcrop, possess little solidity, and present, to some extent, the appearance of mineral charcoal. These samples, besides their liability to contain sulphur, possess little durability in combustion, and but low heat-producing properties. Still other samples, taken from the more solid portions of the seam, present a degree of lustre, hardness, homogeneity and purity, which entitle them to a place in the very first rank of bituminous coals It is evident that our coals ought to be judged from the character of these deeper-seated portions of the seam.

Coal has been mined at several points in the vicinity of Jackson. At Woodville and Barry, the work has been prosecuted with great energy and perseverance. At the latter place, drifts have been carried in from the outcrop. I am informed by Mr. Penny, one of the Directors of the company, that they are now taking out about five hundred tons of coal per month, and that it sells readily at the following prices :

Prices of Stevens' Ridge Coal, per ton.

	COARSE.	NUT.	SLACK.
Delivered on M. C. R. R. Cars,	$2 50	$2 00	$0 70
" M. S. R. R "	3 00	2 50	1 30
" to order in Jackson,	3 50	3 00	1 50
" at the mine,	2 75	2 25	1 50

This coal is said to burn very freely in stoves and grates, and to be free from "clinker." The "Nut Coal" and "screenings" are excellent for making steam, and are used quite extensively by blacksmiths. The gas-producing properties of the coal are good According to a certificate of John Murray, Superintendent of the Jackson City Gas Co., an extract, taken at random from the Register of the works, proves this coal to produce, on an average, 3.83 cubic feet of gas per pound of coal; and as the records were kept while the retorts were in a leaky condition, Mr. Loomis, one of the Directors of the Gas Company, certifies that the real production of gas was not less than 4.20 cubic feet per pound of coal. The gas is very rich—a fact of as much importance as the quantity produced—having from 25 to 50 per cent. more illuminating power than that made from "Willow Bank," and some other Ohio coals. The quantity of lime necessary for purifying the gas, is about two bushels per ton of coal. The yield of coke is said to be about forty bushels per ton of coal, and is of a good quality. With proper ovens, it can be coked to advantage.

According to information from P. E. Demill, Esq., Superintendent of the Detroit Gas Light Co., 6850 lbs. of coal "from Jackson Co." produced 29,400 cubic feet of good illuminating gas, showing a yield of 4.29 feet to the pound of coal. He also obtained from the same quantity thirty bushels of coke,* weighing twenty-nine lbs. to the bushel, the standard weight being thirty-two lbs. to the bushel. This experiment was made in 1857, at a time when the quality of the coal taken out would be likely to yield a lighter coke than the coal at present obtained.

At Woodville, a shaft was sunk about 90 feet, and chambers have been excavated in various directions from the bottom of the shaft. A large quantity of coal has already been taken out.

---

*I cannot avoid thinking Mr. Demill means to say thirty bushels *per ton* of coal used. Mr. Holcroft certifies that he gets forty bushels per ton. It may be added that thirty to forty bushels per ton of coal is the usual yield of coke from the English gas-producing coals (Clegg on Coal Gas, p. 121, &c.) The amount of coke is inversely as the amount of gas.

The Woodville mine was first opened in 1857. It proved, on working, to be located within a small basin about 500 feet in diameter, the rise of the coal to the outer edge being about eleven feet. Within this basin the coal is intersected by numerous faults, which cause a deterioration of its quality for several feet on each side. On extending the working of the mine, however, beyond the rim of the basin, the seam of coal is found to have greater regularity, compactness and purity. In consequence of the peculiar locality of the mine, the company have been obliged to deliver a grade of coal somewhat impure, but the present workings are bringing out an article of improved quality.

The coal of this mine is used with success both for domestic and steam purposes. It ignites freely in an open grate, emits a cheerful flame, and produces as much heat as any other bituminous coal. It is used for heating the Insane Asylum at Kalamazoo. It is also declared to be a superior article for generating steam. The screenings and refuse are used for engine fuel at the mine, and are taken by blacksmiths for their use to the distance of twenty miles north and south of the railroad.

This coal makes a good coke for locomotives, malt houses, &c, but for want of facilities for making it in large quantities, the coke has not yet been tested in furnaces for the manufacture of iron. There can be no doubt that for gas purposes this coal would be found similar to the Stevens' Ridge coal.

I am under obligations to John Holcroft, Esq., for particular information respecting this locality.

Mr. Alexander McArthur has taken large quantities of surface coal from an outcrop near Corunna. This coal has long been in request for blacksmithing purposes. Recently Messrs. Frazer and Stanton have sunk a shaft at a point where the coal lies several feet from the surface, and below the thinned prolongation of the Woodville sandstone. Accounts state that they are now daily sending several tons to the Detroit market.

The gas producing properties of the Corunna coal were also

tested by the Detroit Gas Light Company, in 1857; but as any coal taken out at that time, necessarily came from the immediate outcrop of the seam, the result of the trial would throw no light on the permanent qualities of tl e seam.

It is obvious that Michigan has a very great interest in the development of this resource. Her forests are rapidly receding before the axe, and the demand for coal is yearly increasing. The amount of coal introduced into Detroit from Cleveland and Erie during the present year is stated to be about 26,000 tons; and this has not equaled the demand. The consumption, nevertheless, in consequence of the supply held over from last year, has been 33⅓ per cent. greater than for 1859. This amount, at $5 per ton, gives $130,000 as our annual tribute to the coal mines of Ohio and Pennsylvania, through the Detroit market alone.

MATERIALS FOR PAINTS.

Ochre beds are found in Jackson county, embraced in the Woodville Sandstone.

At several localities, ochreous deposites from springs have been found existing in such quantity as to justify attempts at establishing a business. The N. E. ¼ of Sec. 21, Sharon, Washtenaw county, on the land of J. Townsend, is one productive locality. The deposite covers about 16 square rods, and is seven feet deep. Another deposite covers three acres.

An extensive deposite of a black substance, supposed to be oxyd of manganese, occurs on the same farm, at the depth of two feet beneath a bed of peat. It is 14 inches thick, and covers an area of two or three acres. Mr. L. D. Gale, of Grass Lake, has used this paint quite extensively on carriages.

Ferruginous and chocolate colored shales occur at numerous localities in the coal measures, and might undoubtedly be made to afford a good mineral paint. A paint of this kind has been used for outside work at Lansing, and has stood well for two years.

GYPSUM.

So much has already been said of the geographical and geological position of the gypsum of our State, that I only add a few memoranda.

The following analyses were performed, at my request, by Prof. L. R. Fisk, of the Agricultural College:

	Grand Rapids Gypsum.	Ohio Gypsum.
Water,	20.8445	20.8631
Silicic acid,	Trace.	.0235
Alumina and Oxyd of Iron,	.5354	.7626
Sulphuric acid,	46.2257	45.8303
Lime,	32.0385	31.5628
Potassa,	.2115	.2676
Soda,	.0140	.0944
Chlorine,	.0078	.0050
Total,	99.8774	99.4093

The above statement does not exhibit at a glance, the comparative purity of the two products; we therefore calculate the following further results:

	Grand Rapids.	Ohio.
Lime, as above,	32.0385	31.5628
Sulphuric acid required for this,	45.7696	45.0897
Water required for these two,	20.5962	20.2903
Total hydrous gypsum,	98.4043	96.9428
Excess of sulphuric acid,	.4561	.7406
Excess of water,	.2483	.5728
Other constituents,	.7687	1.1531
Total as before,	99.8774	99.4093

It thus appears that the sample of Grand Rapids gypsum analyzed, contained only 1.5957 parts in 100, of impurities, and the Ohio gypsum only 3.0572. Of these impurities, however, the sulphuric acid, potash and soda, are at least equally valuable with pure gypsum. These ingredients amount to .6816 in the Grand Rapids sample, and 1.1026 in the Ohio sample, leaving for the residual, worthless constituents of the former .9141 per cent. and of the latter 1.9546 per cent.

The following analysis is said to have been made by Dr. S. P. Duffield, of Detroit:

	Grand Rapids.	Ohio.
Water,..................................	19.00	20.70
Lime,..................................	32.67	32.27
Sulphuric acid,..........................	44.44	45.95
Organic matter and loss,.....................	3.89	1.08
	100.00	100.00

Here, on the contrary, the Ohio gypsum contains the most sulphuric acid. It appears, however, that the quantity stated for the Grand Rapids sample, is not sufficient to neutralize the lime by 2.23.

By unfair selection of samples, such analyses may be made to show anything. The gypsum as it finds its way to the market is a mixture of different grades. The only true test would be an analysis of average samples taken direct from the market, not picked for the occasion. The samples sent Prof. Fisk were nearly the best of each. It is but justice to say, however, that a large proportion of the Grand Rapids gypsum, is equally fine with the specimen analyzed.

I have been unable to ascertain the extent of the plaster business at Grand Rapids during the past year. While this report is passing through the press, I am furnished by Mr. Freeman Godfrey with some interesting facts relative to the operations of a new company, and I desire to stimulate other companies to a greater attention to their true interests, by making the following brief mention of the "Florence Plaster Mills," near Grand Rapids.

Last October Mr. Godfrey purchased 103 acres of land upon Plaster Creek, upon the south side of Grand River, and at once began extensive preparations for the quarrying and grinding of gypsum. At present the mill which has been erected is turning out 40 tons of ground plaster per day, and Mr. Godfrey intends putting in another run of stone next summer, and erecting a building for the manufacture of stucco. The quarry is situated in close proximity to the mill. The plaster is reached by stripping, 15,000 cubic yards having been already excavated, and 1,200 tons of plaster taken out from a bed 12 feet thick. At the present time from 60 to 100 tons are quarried per day.

The amounts of gypsum received at Detroit during the past year are as follows:

	Tons.
From Grand Rapids,................................	6,030
" Sandusky, ...	4,661
Total,................................	10,691

Unless gypsum should be discovered in Monroe county, the region along the Southern railroad will continue to be supplied from Ohio; but the greater portion of the State will soon be supplied with gypsum of our own production.

SALT.

The manufacture of salt is rapidly assuming a great degree of importance to our State. If the geological indications on which I found my opinions are not fallacious, we have the most magnificent saliferous basin upon the continent, east of the Mississippi. As might be expected, too, the strength of the brine is proportioned to the extent of the basin.

I omit any historical notices of the rise and development of this interest from the time when the State commenced legislating on this subject, in 1836, down to the present. Many of the disappointments heretofore experienced, might have been avoided by an observance of such practical suggestions as are subjoined:

1. The occurrence of a salt spring is a fact of no consequence whatever, except in connection with all the other geological facts.

2. Brine is found issuing at the outcrops of the coal measures, the Gypseous Group, the Napoleon Group, the Marshall Group and the Onondaga Salt Group. In Ohio, it also issues from the Coal Conglomerate, the Hamilton Group and the Hudson River Group.

3. Only two of these groups will be found, in our State, to

produce brine of sufficient strength for manufacturing purposes; and at present only the Gypseous Group is known to do this.

4. Before deciding on the indications of a salt spring, therefore, it is necessary to know from what geological formation it issues. Here the elaborate investigation of the order and distribution of our strata, finds one of its applications.

5. Before the origin of the brine can be known, we must ascertain whether it flows out horizontally at an outcrop, or rises vertically through fissures in strata overlying the salt rock. A fundamental mistake, committed in the early explorations for salt, grew out of the assumption that the brine of our springs generally rises through fissures, and may be sought by boring in the vicinity of the springs.

6. Most of our springs issue at *outcrops* of saliferous strata; so that the moment we begin to bore in such situations, we find ourselves *beneath* the source of the salt.

7. The source of the salt must be sought by traveling from the spring toward the center of the basin, when, by boring down, the brine may be expected in increased strength and quantity.

8. Our saliferous basin extends from Grand Rapids to Sanilac county, and an unknown distance toward the north. Within this basin, the area covered by the Coal Measures may be taken as the area underlain by saliferous strata of maximum productiveness.

A great deal of enterprise has been manifested in the establishment of the salt manufacture at Grand Rapids, and a fair degree of success may yet be anticipated. This location is, however, within three or four miles of the outcrop of the saliferous strata, and I have all along thought and stated that the prospects were less encouraging than they would be farther within the basin. The salt bearing strata lie here about 200 feet from the surface; but those who have been engaged in this enterprise have been loth to shake off the old illusion that the great reservoir of the salt lies at the depth of six or eight hundred feet. They have, therefore, in nearly every case, persisted in going down after the "lower salt rock" So far as I know, the uniform result has been a failure; though these explorations have added much to our knowledge of the geology of the State. It will be understood, nevertheless, that by boring sufficiently deep, the Onondaga salt group would be reached, and strong brine *might* rise to the surface. This formation lies about 350 feet below the bottom of Lyon's well.

A company whose efforts were guided by James Scribner, Esq., engaged in the first practical attempt to resuscitate the salt manufacture at Grand Rapids. A well was commenced Aug. 12th, 1859, and finished Oct 14th, being 257 feet deep, and extending 56 feet into the Napoleon group. The well at this time was discharging about 200 gallons of water per minute, of such strength that, according to Mr. Scribner, 224 gallons would produce a bushel of salt. A sample of the brine taken at this time was analyzed by Prof. Fisk, with the following result:

Specific gravity	1.01752	
Fixed constituents,	2.33385	per cent.
Carbonate of Iron,	0.00145	"
" Lime,	0.00473	"
" Magnesia,	0.00084	"
Free carbonic acid,	0.00603	"
Silicic acid,	0.00025	"
Sulphate of Lime,	0.13120	"
Chlorid of Calcium,	0.27641	"
Chlorid of Magnesium,	0.07196	"
Chlorid of Potassium,	0.01561	"
Chlorid of Sodium, (Salt,)	1.73696	"
Loss,	0.08841	"
	2.33385	"

The above amount of solid constituents, if all salt, would require 290 gallons to the bushel. The actual per centage of salt found, would require 392 gallons to the bushel. The impurities are about 26 per cent of the solid constituents. Mr. Scribner subsequently passed a tube to the bottom of the well, so as to eliminate the fresh water, and by this means, obtained a brine of considerably greater strength. In February of this year, I found it standing 5° Beaume, (20° Salometer,) at the temperature of 50° Fah.

In the meantime, borings had been undertaken by the Grand River Salt Co., (Powers, Martin and Leonard,) and by Mr. R. E. Butterworth. The former, on the west side of the river, nearly opposite Scribner's, at the depth of 140 feet, found themselves in possession of brine which stood at 100° of the Salometer. This, however, did not rise to the surface, and the supply was found to be limited. According to information received from Mr. Martin Metcalf, the phenomena presented by this well are as follows:

"When we first put in the pump, only about two quarts per minute were furnished, ranging 80° to 100°. *Now*, we can pump *one gallon in* 16 *seconds*, for 2½ minutes, before we feel a tendency to vacuum; after which, we get 1 40-100 gallons per minute. Now, if we let the well rest five minutes, we can again pump one gallon every 16 seconds, for 2½ minutes, as before, when we find a vacuum creating; and afterwards, if the pumping is continued, we can obtain for half an hour, at least, 1 40-100 gallons per minute. I am told that 1½ gallons per minute have been obtained for several hours together—in fact as long as they have continued pumping—salometer ranging from 80° to 86°."* These phenomena are probably attributable, as Mr. Metcalf suggests, to the existence of a cavity holding about 11½ gallons.

Mr. Butterworth's well attained a depth of about 500 feet. The flow of water from the surface was immense, being not less than 300 gallons per minute. By means of an ingenious arrangement for stopping off the fresh water, Messrs. Metcalf and Butterworth found the brine, at 325 feet, to possess a strength of 22°; and they succeeded in obtaining a constant flow from the top, of one gallon per minute, of the strength of 20°, which would require about 131 gallons for a bushel of solid ingredients.

During the past season, two other wells have been bored.

*Letter dated March 15th, 1860

That of the Indian Mill Creek Co., in the immediate vicinity of Powers and Martin's, was carried to a depth of 434 feet. At 214 feet, brine was found at 19°, in the gypseous group, and at 295 feet, near the bottom of this group, at 27°. No increase was gained in boring the next 139 feet.

Mr. Windsor's well is located three or four miles further north. It has been carried to the depth of 466 feet. The strength of the brine is stated to have increased somewhat after entering the sandstones beneath the gypseous group. At 391 feet, (in the Marshall sandstone,) the salometer stood 1°. The strength of the overflow was 24°, at the rate of about 35 gallons per minute.

Mr. Taylor's well is located very near Scribner's, but I have received no data relating to it.

The manufacture of salt has commenced at Scribner's, Windsor's, and the Indian Creek wells. Scribner erected a brush house, or rather two of them, 12 feet apart. Each house is 100 feet long, 30 feet high, and 7 feet wide. The brine is first passed into a vat holding 32,000 gallons, from which it is pumped by water power to the top of the brush house. From here it falls slowly through six tiers of brush, resting on frames, to the bottom. Thence it flows again into the tank, to undergo the same operation. According to Mr. Scribner, one passage through the brush house in favorable weather strengthens the brine from 26° to 37°. During the process much of the iron is precipitated.

From the vat, the concentrated brine is conveyed to two vats, at the kettle house, each holding 8,000 gallons. Here a little lime is added. From these vats it is conveyed in logs to the 50 kettles. After boiling some time in the seven front kettles on each side, the brine is transferred to four vats, each 6 by 8 by 2 feet, where it is allowed to stand four hours, and precipitate a white substance, which is probably gypsum. From here it is conveyed into the back kettles, and the evaporation continued. As the salt falls down, it is skimmed into baskets and drained. The article manufactured by this process is white and beautiful. After standing some months, a slight deliquescence is perceived, but not as great as upon most of our commercial salt. Experiment shows it to be perfectly free from gypsum; though, of course, rigorous analyses is requisite to fully test its purity. I am not apprised of the results of experiments on its preservative qualities. Two hundred and thirty barrels have been manufactured.

At the Indian Creek Co.'s works the evaporation is conducted in large sheet-iron pans. At Windsor's, both pans and kettles are employed.

The salt business of the Saginaw valley was commenced by the East Saginaw Salt Company, who bored a well about three-fourths of a mile north-east of the village on the bank of the river. This well penetrated the saliferous beds between 464 and 627 feet, and ended at 669 feet. The strength of the water at different depths was as follows: At 70 ft., 1°; at 102 ft., 2°; at 211 ft., 10°; at 293 ft., 14°, discharging 80 gallons per minute of a temperature of 50° Fah., and rising 14 feet above the surface; at 487 ft, 26°; at 516 ft, 40°; at 531 ft., 44°; at 559 ft., 60°; at 569 ft., 64°; at 606 ft., 86°, with a temperature of 54° Fah., and at 639 ft, 90°, or ten degrees short of saturation.

A sample of this water at 64°, from the depth of 575 ft., was examined by Dr. Chilton, of New York, with the following result:

Solid residuum in one wine pint,................	1155	grs.
Chlorid of sodium (common salt,)...............	1014.57	"
Specific gravity,..............................	1.110	

A sample at 86°, from the depth of 617 ft., was examined by Dr. J. G. Webb, of Utica, N. Y., with results as follows:

Chlorid of sodium in one wine pint,...............	1416 grs.
Other chlorides,..................................	32 grs.*

Subsequently a more detailed analysis of the brine was made by Dr. Webb, the results of which are given below:

*There must be an error in this amount, as this brine standing at 86° should have about 1800 grs. of solid matter to the pint.

	Per Cent.
Chlorid of sodium,	19.088
" " calcium,	.537
" " magnesium,	1.241
Sulphates of lime and magnesia,	.225
Total solid matter,	21.091

The following analysis was published by Prof. Douglass April 16th, 1860, said to have been performed upon water from the salt well of "Mr. Waldron, of Saginaw," but according to Mr. Waldron, of East Saginaw, taken from the well of the above company:

Specific gravity,	1.170	
Saline matter,	22.017	per cent.
Chlorid of sodium, (salt,)	17.912	"
Sulphate of lime, (gypsum,)	.116	"
Chlorid of lime, [C. of calcium?]	2.142	"
Chlorid of magnesia, [C. of magnesium?]	1.522	"
Carbonate of iron,	.105	"
Chlorid of potassium,	.220	"
Water,	77.983	
	100.000	

The chlorid of calcium given here is four times the amount found by Dr. Webb.

In May, 1860, another and more accurate analysis of this brine was made by Dr. Chilton, with the following results:

Specific gravity at 60° Fahrenheit, 1.177

In 100 parts of brine, Salometer 90°, are found,

Chlorid of sodium,	16.8710
" " calcium,	3.2873
" " magnesium,	1.7743
Bromid of sodium,	.0401
Sulphate of lime,	.0982
Carbonate of lime,	.0500
Silica and alumina,	.0245
Carbonate of iron,	.0116
Water,	77.8430
	100.0000

In 100 parts of dry solid matter, there are,

Chlorid of sodium,	76.143
Other substances,	23.857
	100.000

The total per centage of solid matter is 22.157.

In one wine pint there is, of solid matter,

Chlorid of sodium,	1229.72 grs.
Other saline matters,	385.30 "
	1615.02 "

This well furnishes about 13,000 gallons of brine in 24 hours, ranging in strength from 75° to 80° by the salometer. According to Dr. H. C. Potter, Superintendent of the works, the brine is treated in the following manner: "We get a deposite of iron in our settling vats, first, by putting the brine into them heated, (running it through a heater,) and, second, by using on each 27,000 gallons a pailful of lime. We are trying experiments to settle with other materials. In the kettles we used alum for cleansing for a time, but recently, and since cold weather, we have used nothing. The chlorides can only be removed by bailing out the residuum, after say the 5th to the 8th drawing of salt, when the bitter water accumulates to such an extent as to act on the iron of the kettle, and rust the brine and the salt. This course, of throwing out the bitter water, is adopted in Kanawha, Va., and Pomeroy, Ohio, where the brine resembles ours in chemical composition, and though an expensive one in loss of brine, seems the only one that is practicable. * * * The impurities remaining in our salt, after having been drawn from the kettles, are removed by drainage, being liquid almost entirely. This thorough drainage is *the* essential point in our manufacture."

This company are engaging vigorously in the manufacture of salt, both by solar evaporation and by boiling. They have 20 covers, 16 feet square, for solar evaporation, and 100 kettles for artificial heat. They have produced to the date of this report about 4,500 barrels of salt, and are making 600 to 800 barrels per month. The salt is put up in extra quality of white-oak barrels, costing 28 cents each. Fuel, of hard and soft wood mixed, costs, delivered at the works, about $1 38 per cord.

The quality of the salt produced is unsurpassed, either in chemical purity or preservative qualities. Several of the most extensive fishermen upon the lake shore having given it a thorough trial, pronounce it "more economical, (in quantity required,) safer and better than the Onondaga fine salt." It is equally commended by butchers. For butter it has been tested both in our own State and in Orange county, N. Y., and pronounced not at all inferior to the famous Ashton salt.

This company have sunk another well during the past summer, under the same roof, to the depth of 806 feet. Though the tubing has not been inserted, the indications are that a larger supply of strong brine has been obtained. The only brine drawn up stands at 90°.

Numerous other enterprises have been started along the Saginaw river, of which one is at Saginaw City, one at Carrolton, one at Portsmouth, and one at Bay City. According to information received from Wm. Walker, the strength of the brine at various depths, in the Bay City well, was as follows: At 223 feet, 5°; at 229 ft., 8°; at 235 ft, 12°; at 245 ft., 14°; at 256 ft., 16°; at 270 ft., 18°; at 273 ft., 19°; at 434 ft., 20°; at 438 ft., 42°; at 444 ft., 60°; at 450 ft., 70°; at 480 ft., 78°; at 487 ft., 85°; at 490 ft., 90°. From this point to the depth of 513 feet, it varies between 88° and 92°.

At the date of the printing of this report, the following parties have either completed salt borings or have them in progress. For the statements of outlay to Jan. 1st, 1861, and estimates to June 1st, I rely upon an editorial article in the Detroit *Tribune*:

I.—WELLS ON THE GRAND RIVER, (AT AND NEAR GRAND RAPIDS.)

	Depth. Feet.
1. Grand Rapids Salt Manufacturing Co., (Scribner & Co.,)	410
2. Grand River Salt Manufacturing Co., (Ball, Clay & Co.,)	402
3. R E Butterworth,	500
4. Indian Mill Creek Salt Manufacturing Co., (Powers & Martin,)	450
5. J. W. Windsor,	446
6. Taylor,	402

II.—WELLS ON THE SAGINAW RIVER.

	Depth. Feet.	Outlay to Jan. 1.	Estimated to June 1.
7. East Saginaw Salt Manufacturing Co., (E. Saginaw,) 1st well,	669	$25,000	$40,000
8. Do., 2d well,	806		
9. Saginaw City Salt Manufacturing Co, (Saginaw City,)	600	9,000	12,500
10. Hall, Gilbert & Co., (Florence,)	350	2,000	5 000
11. Ward, Curtis & Co., (Carrolton,)	560	7,000	10,000
12 E. Litchfield & Co., "	680	4,000	10,000
13. G. A. Lathrop & Co, "	(Commencing.)	5,000	10 000
14. Portsmouth Co., (Portsmouth,)	667	7,500	12,500
15. Bay City Salt Manufacturing Co., (Bay City,)	542	8,000	12,500
16. New York Salt Manufacturing Co., (East Saginaw,)	(Commencing.)		
17. Saginaw and Buena Vista Salt Co.,	"		

It is proposed to continue the boring in Taylor's well, at Grand Rapids, until the *Onondaga Salt Group* is reached.

The East Saginaw Co. are manufacturing 40 to 50 barrels of salt per day, and on getting their second arch of kettles in operation expect to produce about 100 barrels per day. This "Company have received the diploma of the New York State Agricultural Society, and the prize medal from the Mechanic's Institute of Chicago."

The Saginaw City Co. have 60 kettles on hand, and are proceeding with vigor.

It cannot be denied that the prospects of the ultimate success of the salt manufacture in Michigan are exceedingly encouraging. Aside from the unparalleled strength of the brine of the Saginaw valley, the position, surrounded by forests, which must cheapen to the last degree the cost of barrels and fuel, and upon the immediate shore of navigable waters stretching from Oswego to Chicago, is such as to enable us to compete successfully with any other source of supply to the western and north-western States.

As to the actual cost of producing a barrel of salt at either of the points at which the manufacture has been commenced, I am not in possession of the data to enable me to speak definitely. At Saginaw, as I am authentically informed, wood of mixed quality, (*i. e.* "hard" and "soft,") can be delivered for $1 38 per cord. At Syracuse, experiments have shown that one cord of hard wood will produce, in blocks of 50 or 60 kettles, an average of about 53 bushels of salt. Assuming, as is done at Syracuse, that two cords of hard wood are worth three of soft, the cost of hard wood at Saginaw should be $1 656. Reckoning 53 bushels to a cord of wood, this would make the fuel cost at Saginaw $0 031 per bushel, or $0 155 per barrel of salt. If one block of kettles is capable of producing but 40 barrels of salt per day, and the services of six men, at $1 00 per day, are required to attend them, the element of labor entering into the cost of a barrel is $0 15. At Syracuse barrels cost 25 cents each, and I see no reason why they cannot be produced for much less than this at Saginaw. I am informed, however, that the lowest bids offered are 27 cents per barrel. It is admitted, however, that this is for a superior article. At the works of the East Saginaw company, where it is stated 40 barrels per day are now manufactured, it is reliably announced that not more than $25,000 have been expended in boring two wells, the largest and deepest of which has not yet come into use. Assuming that one half this sum has been expended in boring the well now in use, and that capital is worth 10 per cent, the annual interest on the investment is $1,250, or $4 166 per day, or $0 104 per barrel of salt produced. Should the wear and tear of fixtures and apparatus amount to 5 per cent.

more, this item would add $0 052 to the cost per barrel. The cost of packing is stated to be 2½ cents per barrel at Syracuse, and it could not be greater at Saginaw.

Bringing together now these various items, we find the cost of a barrel of salt at Saginaw to stand as follows:

Fuel, hard and soft equally mixed,....................	$0 155
Labor of six men, at $1 00 per day,.................	0 150
Barrel of superior quality,..........................	0 270
Packing, ...	0 025
Interest at 10 per cent. on $12,500,.................	0 104
Wear and tear at 5 per cent.,........................	0 052
Total,.................................	$0 756

Aside from the cost of superintendence and incidentals, it does not appear how the above aggregate can be materially increased when the business is once fairly established. At the same time it must be admitted that it is rather early in the history of the enterprise to venture upon calculations as to the ultimate minimum cost of the manufacture. As an existing fact it should be borne in mind that, aside from the greater expense attending the commencement of any manufacture, and that which is always incident to manufacturing on a small scale, the chlorid of calcium which exists in considerable quantity in all our brines, will materially enhance the cost of production until some cheap method is discovered of eliminating it by chemical precipitation. With whatever confidence, therefore, we may speak of the ultimate prosperity of this manufacture in our State, it should not be forgotten that the enterprise is still in its infancy; and, in view of the powerful competition arrayed against it, still needs the fostering care of the government to a liberal extent.

Whether such liberality ought to extend to a continuance of the existing bounty on the manufacture of salt, depends upon various considerations, which the legislature alone will be competent to estimate as a whole. In the meantime it may not be amiss to offer the following suggestions, derived from geological data. I purposely ignore the questions whether it is morally honest to discontinue the bounty at the present juncture, and whether the State is pecuniarily able to continue any bounty, as these are not geological questions:

1. Whatever may be the state of the salt enterprise at Saginaw, the business is *not* established at any other point.

2. Though we believe strong brine may be procured throughout the center of the State, this belief is purely a geological inference. The public interest would be vastly promoted by bringing this theory to the test of experiment.

3. Even supposing it certain that the *Michigan Salt Group* will prove productive throughout the center of the State, there is still another vast salt basin which has never been explored, within our limits. This is situated about 800 or 900 feet below the other basin, and literally underlies the entire peninsula. Its margin rises to the surface at Mackinac on the north, Milwaukee on the west, Sylvania, Ohio, and Monroe county, Mich., on the south, and Galt, in Canada West, on the east. It is the source of all the brine worked at Syracuse and vicinity, in the State of New York. There are some indications that the great basin formed by these rocks in Michigan is also filled with brine. Suppose this to be the case. The result would be that every county in the peninsula might become a salt producing county. If it is not desirable to restrict the benefits of the establishment of this manufacture, the State has an interest in stimulating the exploration of these lower rocks. The offer of a bounty would cost the State nothing unless the attempt should prove successful. If successful, the payment of the bounty would prove one of the best investments the State ever made.

4. Should it not, after all, appear to be good policy to stimulate researches by the offer of bounties, there are still other methods by which the spirit of enterprise now awakened may be seconded, unless indeed all idea of public encouragement to the development of our State resources is to be entirely abandoned. The discovery of some economical means for the separation of the chlorid of calcium, which constitutes the principal

difficulty in the working of our brine, is an object which ought not to be left to the chances of private enterprise. The policy recognized, and the experience gained in all similar cases in the history of the past, both advise the setting apart of a special sum as a proffered reward for successful discovery in this direction. Such reward should be open to universal competition. If success were not attained, no expense would accrue to the State. If success were reached, millions of dollars would be added in a day to the wealth of our people.

I must be pardoned for making the following further suggestion:

If the State of Michigan contains a population of 750,000, the total annual consumption of salt, estimating at the rate of 45 lbs. per capita, is about 602,000 bushels, or 120,420 barrels. During the past year the average price of salt delivered in Detroit has been about $1 50 per barrel. At this rate the annual contribution of Michigan to the business of other States (except the small amount paid for freight on salt carried in Michigan vessels) is $180,630. Every political economist must recognize the desirableness of retaining this expenditure within our own State. The amount which the State could afford to expend to effect this object would be the annual interest of the money of which the State is thus deprived of the use. That money is the whole amount of the profit to the manufacturer and dealer until the commodity reaches our own borders.

Having presented as extended a statement as seems immediately necessary, of the local details connected with the salt enterprise in our State, it may be useful, in view of the general interest felt in this new branch of industry, to append some general and comparative statements for the purposes of reference and comparison.

As the addition of common salt to pure water increases its weight, bulk for bulk, it follows that the strength of any brine may be known by comparing its weight with that of the same bulk of pure water. This comparative weight is its *specific gravity*. The most exact method of ascertaining the specific gravity of brine is by weighing a given bulk of it with a delicate balance. The most convenient method, however, is by means of a *hydrometer*, and this is sufficiently accurate for most practical purposes. A hydrometer is an instrument generally made of glass, in the shape of a tube closed at both ends with a large bulb blown in it, and a weight attached at the lower end. The tube is graduated above the bulb, in such a manner that when the instrument is placed in pure water it sinks to a mark designated 0, and when placed in a liquid heavier than water, it sinks to some mark below the first, against which is the figure which designates the true specific gravity of the fluid. In *Beaume's* hydrometer, which is the one most used for general purposes, the figures on the scale do not designate the specific gravity directly. The scale is graduated from 0 to some arbitrary point which reads 30° or 40°—the intervening space being equally, or nearly equally, divided, so that the specific gravity can only be known from it by a calculation.

The hydrometer, however, which is most convenient for experiments with brine, is the one which marks 0° when immersed in pure water, and 100° when immersed in saturated brine. This instrument is called a *salometer*. The number of degrees indicated upon the salometer, therefore, is the per centage of saturation possessed by the brine. We may speak of 25° on the salometer or 25 per cent. of saturation.

It must be distinctly understood, that 25 per cent. of saturation does not mean that 25 per cent. of the brine is composed of salt. Twenty-five per cent. of salt produces 100 per cent. of saturation—and this happens to be 25° on Beaume's hydrometer.

It is apparent, therefore, that the specific gravity of a brine, the readings of the hydrometer and salometer, and the per centage of salt. are all different expressions for the same thing, which may also be expressed by the number of gallons of brine required for a bushel of salt of 56 lbs. As it is often desirable to convert these expressions into each other, I have calculated the subjoined table which, it is believed, will be found useful, and sufficiently accurate.

It must be borne in mind that the calculations are based upon the supposition that the brine contains no foreign constituents; but as all natural brines do contain varying amounts of foreign constituents, it follows, first, that the total amount of solid matter does not bear the same ratio to the density, as if the brine were pure; and secondly, that the amount of salt may be quite a different thing from the amount of solid constituents, which alone determines the density or specific gravity.

It must also be borne in mind that brines of the same strength possesses different densities depending upon their temperature—the density rapidly diminishing as the temperature rises. It is consequently necessary to experiment on brines at a uniform or standard temperature. The ordinary standard temperature for hydrometrical operations is 60 degrees, Fahrenheit's thermometer, but the standard temperature at the Onandoga salines, is 52°, that being the natural temperature of the brine as it issues from the well. As the natural temperature of Michigan brines, obtained from ordinary depths, would be nearly the same, 52° might have been adopted as the standard in the following calculations. The results, however, would not have been practically different from those given.

Constants, useful for reference, a portion of which are deduced from the subjoined calculation, and others the data upon which the calculation is based:

1. Specific gravity of pure water,....................1.
2. Specific gravity of common salt, according to Ure, 2.0 to 2.25 (mean),...........................2.125
3. Specific gravity of saturated brine,................1.205
4. According to Ure, 100 parts of water dissolve, at 62½° Fah., 35.88 parts of salt.
5. One bushel of salt=9.3 gallons, wine measure; dissolves in 16.8 gallons of water, making, without allowing for condensation, 26.1 gallons of brine.
6. One hundred volumes of the constituents of a saturated solution of salt, become, by condensation, a little less than 96 volumes, (Ure.)

7. One cubic foot of saturated brine weighs 85 lbs.
8. One bushel of salt weighs 56 lbs.
9. One wine pint contains 26 625 cubic inches.
10. One wine pint of distilled water weighs 7288.975 grains.
11. Every .001 variation in specific gravity corresponds to about .25 gallon of brine required for a bushel of salt.
12. One degree of Beaume=4° Salometer, approximately.
13. Specific gravity $=\frac{152}{152-\text{Deg. Beaume.}}$
14. Gallons of brine to a bushel of salt $=\frac{2603.88}{\text{Deg. Salom.}}-4.454.$

Let s=per centage of salt in any brine,
g=specific gravity of the brine,
B=its density, by Beaume's hydrometer,
S=its percentage of saturation, by the Salometer,
G=number of gallons required for 1 bu. of salt,

Then the value of each of these quantities may be expressed in terms of each of the others, as shown by the following twenty equations:*

1. $G=\frac{2603.88}{S}-4.454$ When S=o, G=∞
2. $G=\frac{677.008}{B}-4.454$ " B=o, G=∞
3. $G=\frac{670.218}{s}-4.454$ " s=o, G=∞
4. $G=\frac{4.454}{g-1}$ " g=1, G=∞

---

*As the standard bushel of salt weighs 56 lbs.,

$$\frac{5600}{s}=\text{pounds of brine required for 1 bu. of salt.}$$

And since one gallon of distilled water weighs 8.355 lbs.,

$$G=\frac{5600}{8.355\,g\,s}=\frac{670.257}{g\,s} \qquad (1)$$

But the value of s, or the per centage of salt in the brine, may be expressed in terms of the specific gravity of the brine. For, the specific gravity of the brine is its weight divided by the weight of the same bulk of water. Making no allowance for condensation of the aggregate volume of the constituents, this would be

$$g'=\frac{100}{w+\frac{s}{2.125}} \qquad (2)$$

In which w is the per centage of water in the brine, and 2.125 is the mean specific gravity of salt. But experiment shows that some degree of condensation always takes place; and it seems obvious that the amount of this condensation must be a direct function of the per centage of salt in the solution. The data at command, however, do not seem to be consistent with this theory nor with each other.

According to Dr. Ure, 100 measures of the constituents of a saturated solution, make a little less than 96 measures of the brine; and this brine contains 25.5 per cent. of salt. Now, as the specific gravity of this saturated solution is, by the same authority, 1.1962, we may calculate what would have been the specific gravity without condensation. This would be

$$\frac{1.205+96}{100}=1.1568$$

5. $S=\frac{2603.88}{G+4.454}$ " $G=\infty$, $S=0$

6. $S=3.846\,B$ " $B=0$, $S=0$

7. $S=3.885\,s$ " $s=0$, $S=0$

8. $S=585.516-\frac{585.516}{g}$ " $g=1$, $S=0$

9. $B=\frac{677.008}{G+4.454}$ " $G=\infty$, $B=0$

10. $B=.26\,S$ " $S=0$, $B=0$

11. $B=1.01\,s$ " $s=0$, $B=0$

12. $B=152-\frac{152}{g}$ " $g=1$, $B=0$

13. $s=\frac{670.218}{G+4.454}$ " $G=\infty$, $s=0$

But knowing the mean specific gravity of salt to be 2.125, we may also calculate the specific gravity of the saturated solution (without allowance for condensation) from the per centage of salt, by means of formula (2). This gives

$g'=1.1560$

It is evident, therefore, that Dr. Ure's value of the condensation is too great, or else his per centage of salt in satur ted brine is too great. But that per centage is less than given by most other authorities, while by my own experiments upon commercial salt, it amounts to 26.595.

Again, according to the experiments of MM. Francœur and Dulong, when a brine contains 10 per cent. of salt, its specific gravity is 1.0735; and when it contains 15 per cent., it is 1.1094. Now if we assume 10 for the per centage of salt in Eq. (1), we get

$g'=1.0559$, instead of 1.0735.

If we assume 15 for the per centage of salt,

$g'=1.0862$, instead of 1.1094.

The increased specific gravity due to condensation in the first case, is .0186=.186 per cent. of 10, the per centage of salt.

In the second case, it is .0232=.155 per cent. of 15, the per centage of salt.

Further, in the case of saturated brine, it is .049=.191 per cent. of 25.5, the per centage of salt. The first and last values are sufficiently consonant, but not so the second. The mean of the first and last is .188 per cent. Assuming this

$g=g'+.00186\,s$

Substituting the value of g' from Eq. (2), we might thence deduce s in terms of g.

Another view may be taken of this subject. It is evident that we may regard all the condensation as taking place in the salt; and the result will be the same if we imagine it to take place before the solution. We may then proceed to calculate what value of the specific gravity of the salt would be requisite in order to produce, without further condensation, a brine of a given specific gravity, and containing a given per centage of salt.

If in (2) we make $g'=1.0735$, $w=90$, $s=10$ and $2.125=x$, we get

$x=3.186$.

If in (2) we make $g'=1.094$, $w=85$, $s=15$, and put x for 2.125, we get

$x=2.919$.

If again we make $g'=1.205$, $w=74.5$, $s=25.5$,

$x=2.838$.

These results are but little accordant; and show that the condensation is not proportional to the per centage of salt, or else that errors exist in the data. The mean of the three values is 2.981.

If now in Eq. (2) we substitute 2.981 for 2.125, g' ought to become g, when we should have

$$g=\frac{100}{w+\frac{s}{2.981}}=\frac{100}{100-s+\frac{s}{2.981}}=\frac{150.478}{150.478-s}\quad\text{.......... (3)}$$

Whence, also,

$$s=150.478-\frac{150.478}{g}\quad\text{.......... (4)}$$

14. $s = .257\,S$ " $S=0,\ s=0$

15. $s = .99\,B$ " $B=0,\ s=0$

16. $s = 150.478 - \frac{150.478}{g}$ " $g=1,\ s=0$

17. $g = \frac{4.454}{G} + 1$ " $G=\infty,\ g=1$

18. $g = \frac{585.516}{585.516 - S}$ " $S=0,\ g=1$

19. $g = \frac{152}{152 - B}$ " $B=0,\ g=1$

20. $g = \frac{150.478}{150.478 - s}$ " $s=0,\ g=1$

Further, the number of grains of salt in a wine pint is

$$\text{Salt} = \frac{10968.268\ S}{585.516 - S}$$

From these formulæ the following table has been calculated :

And substituting this value of s in Eq. (1),

$$G = \frac{670.257}{150.478\,g - 150.478} = \frac{4.454}{g - 1} \quad \text{.........} \quad (5)$$

It is often desirable to know G in terms of the degrees of Beaume's scale. This value may be obtained from the equation

$$g = \frac{152}{152 - B}$$

(See McCulloch, Rep. on Sugar and Hydrometers, p. 71) in which B represents the degrees of Beaume's hydrometer expressive of the density of the brine. Substituting this value of g in (5), we get

$$G = \frac{677.008}{B} - 4.454 \quad \text{.........} \quad (6)$$

Since 26° Beaume, or 100° of the salometer, marks saturated brine, it appears that one degree of Beaume equals 3.846 of the salometer; or, putting S for the reading of the salometer

$$B = .26\ S$$

And substituting this value of B in (6), we get

$$G = \frac{2603.88}{S} - 4.454 \quad \text{.........} \quad (7)$$

From which may be calculated a table giving the number of gallons of brine required for one bushel of salt, at every degree of the centigrade salometer.

Although, owing to the inconsistency of the data employed, the foregoing formulæ can give only approximate results, they may be sufficiently accurate for practical purposes; and hence a table has been based upon them.

TABLE giving a comparison of different expressions for the strength of Brine from zero to saturation.

Salometer.	Beaume.	Specific gravity.	Per cent. Salt.	Grains to 1 pint.	Gallons to 1 bushel.	Salometer.	Beaume.	Specific gravity.	Per cent. Salt.	Grains to 1 pint.	Gallons to 1 bushel.
0	0	1.000	0.	0	Infinity	51	13.26	1.095	13.11	1047	46.6
1	.26	1.002	0.26	19	2599	52	13.52	1.097	13.36	1070	45.6
2	.52	1.003	0.51	3[illegible]	1297	53	13.78	1.100	13.62	1092	44.7
3	.78	1.005	0.77	56	863	54	14.04	1.102	13.88	1115	43.8
4	1.04	1.007	1.03	75	647	55	14.30	1.104	14.13	1137	42.9
5	1.30	1.009	1.2[illegible]	9[illegible]	516	56	14.56	1.106	14.39	1160	42.0
6	1.56	1.010	1.54	114	430	57	14.82	1.108	14.65	1183	41.2
7	1.82	1.012	1.80	133	363	58	15.08	1.110	14.91	1206	40.4
8	2.08	1.014	2.06	152	321	59	15.34	1.112	15.16	1229	39.7
9	2.34	1.016	2.31	171	285	60	15.60	1.114	15.42	1252	38.9
10	2.6[illegible]	1.017	2.57	191	256	61	15.86	1.116	15.6[illegible]	1276	38.2
11	2.86	1.019	2.83	210	232	62	16.12	1.118	15.93	1299	37.5
12	3.12	1.021	3.08	229	213	63	16.38	1.121	16.19	1322	36.9
13	3.38	1.023	3.34	249	196	64	16.64	1.123	16.45	1346	36.2
14	3.64	1.025	3.60	269	182	65	16.90	1.125	16.70	1370	35.6
15	3.90	1.026	3.85	288	169	66	17.16	1.127	16.96	1393	35.0
16	4.16	1.02[illegible]	4.11	308	158	67	17.42	1.129	17.22	1417	34.4
17	4.42	1.030	4.37	32[illegible]	149	6[illegible]	17.68	1.131	17.4[illegible]	1441	33.9
18	4.68	1.032	4.63	348	140	69	17.94	1.133	17.73	1465	33.3
19	4.94	1.034	4.88	368	133	70	18.20	1.136	17.99	1489	32.7
20	5.20	1.035	5.14	388	126	71	18.46	1.[illegible]	18.25	1513	32.2
21	5.46	1.037	5.40	408	120	72	18.72	1.140	18.50	153[illegible]	31.7
22	5.72	1.039	5.65	42[illegible]	114	73	18.98	1.142	18.76	1562	31.2
23	5.98	1.041	5.91	44[illegible]	109	74	19.24	1.144	19.02	1587	30.7
24	6.24	1.043	6.17	469	104	75	19.50	1.147	19.27	1611	30.3
25	6.50	1.045	6.42	489	99.7	76	19.76	1.149	19.53	163[illegible]	29.8
26	6.76	1.046	6.68	510	95.7	77	20.02	1.151	19.79	1661	29.4
27	7.02	1.048	6.94	530	92.0	78	20.28	1.154	20.05	1686	28.9
28	7.28	1.050	7.20	551	89.5	79	20.54	1.156	20.30	1710	28.5
29	7.54	1.052	7.45	572	85.[illegible]	80	20.80	1.158	20.56	1736	28.1
30	7.80	1.054	7.71	592	82.8	81	21.06	1.160	20.82	1761	27.7
31	8.06	1.056	7.97	613	79.5	82	21.32	1.163	21.07	1786	27.3
32	8.32	1.058	8.22	634	76.9	83	21.58	1.165	21.33	1811	26.9
33	8.58	1.059	8.48	655	74.5	84	21.84	1.167	21.59	1837	26.5
34	8.84	1.061	8.74	676	72.1	85	22.10	1.170	21.84	1862	26.2
35	9.10	1.063	8.99	697	69.9	86	22.36	1.172	22.10	1888	25.8
36	9.36	1.065	9.25	719	67.[illegible]	87	22.62	1.175	22.36	1914	25.5
37	9.62	1.067	9.51	740	65.9	8[illegible]	22.88	1.177	22.62	1940	25.1
38	9.88	1.069	9.77	761	64.1	89	23.14	1.179	22.87	1966	24.8
39	10.14	1.071	10.02	783	62.[illegible]	9[illegible]	23.40	1.182	23.13	1992	24.5
40	10.40	1.073	10.28	804	60.[illegible]	91	23.66	1.184	23.39	2018	24.2
41	10.66	1.075	10.54	826	59.1	9[illegible]	23.92	1.186	23.64	2045	23.8
42	10.92	1.077	10.79	845	57.[illegible]	93	24.18	1.189	23.90	2072	23.5
43	11.18	1.079	11.05	869	56.1	94	24.44	1.191	24.16	2095	23.2
44	11.44	1.081	11.31	891	54.7	95	24.70	1.194	24.41	212[illegible]	23.0
45	11.70	1.083	11.56	913	53.4	96	24.96	1.196	24.67	2151	22.7
46	11.96	1.085	11.82	935	52.[illegible]	97	25.22	1.198	24.93	2178	22.4
47	12.22	1.087	11.03	957	50.9	98	25.48	1.201	25.19	2205	22.1
48	12.48	1.089	12.3[illegible]	979	49.8	99	25.74	1.203	25.44	2232	21.8
49	12.74	1.091	12.59	1002	48.7	100	26.00	1.205	25.70	2259	21.6
50	13.00	1.093	12.85	1024	47.6	...	....	....	....	....	....

From this table the properties and capabilities of any brine may be ascertained by knowing its strength as shown by the salometer. Suppose for instance the salometer shows 53 degrees. The table shows at a glance that this corresponds to 13.78 degrees of Beaume's hydrometer, a specific gravity of 1.100 and a per centage of 13.62; while a wine pint of the brine would furnish 1092 grains of solid residue, and 44.7 gallons would produce a bushel.* Or suppose the strength of a brine is expressed, as in Dr. Beck's Report, by giving its specific gravity, and we wish to compare the strength as thus stated, with that of another brine given in degrees of the salometer, or the number of grains in a pint, &c. We look in the column of "specific gravity" in the foregoing table and find the number which agrees nearest with the given one, then on the same horizontal line we have all the synonymous expressions for the same strength, and it is seen at once whether the brine with which we wish to make the comparison is stronger or weaker. Or suppose, thirdly, that a land owner desires to know the comparative strength of a brine spring on his premises, while he possesses no instrument for taking specific gravity. Let him evaporate a wine pint and weigh the residue, or take it to the apothecary to weigh; then the number of grains, found in the 5th column of the table, will show him all the equivalent expressions.

In making use of this table it must be remembered that it will prove accurate only for *pure solutions of salt.* In this State the chlorid of calcium which exists to some extent in our brines will cause the table to make a showing a little too favorable. As the per centage of impurities is a variable quantity, it was impossible to make allowance for them in the table. Though we cannot therefore construct a table practically accurate, it was not thought best to discard all attempts at a table. As long as it is thought desirable to use the salometer, it seems to me to be a matter of convenience to have at hand the ready means for converting its reading into the equivalent expressions. This want has been felt by myself, and I have no doubt many others will find the table useful.

TABLE OF ANALYSES OF VARIOUS BRINES.

BRINE.	Specific Gravity.	Chlorid Sodium.	Chlorid Calcium.	Chlorid Magnes.	Sulphate Lime.	Sulphate Magnes.	Carbonate Lime.	Comp'nds of Iron.	Other constituents.	Total Solid Matter.	AUTHORITIES AND REMARKS.
1. Sea Water, (open sea,)	......	2 500	......	0.350	0.010	0 580	0 020	......	......	3 460	Kane, Chemistry, p. 426.
2. " " (East River, N. Y.,)	1 02038	2.030	......	0.226	0.080	0 172	0 122	......	0 050	2 680	Beck, Min., N. Y., p. 112.
3. " " (English Channel,)	1.0274	2.706	......	0.367	0.141	0.230	0 003	......	0 079	3.526	Schweitzer, Encyclopædia Metropolitana.
4. " " (Mediterranean,)	1.0275	2 940	......	0.320	0.130	0.240	0.011	......	0 100	3.741	Usiglio, Encyclopædia of Chemistry.
5. Dead Sea,	1.1300	10.360	3.920	10.246	0.054	......	......	......	......	24.580	Marcet, Phil. Trans. 1807; Lynch, Nar.
6. " "	1.2120	7.039	3.336	12.167	0.052	......	......	......	1.893	24.435	Gmelin & Marchand, Dana's Min. 91.
7. Lake Oroomiah, Persia,	1.1550	18 116	0.148	......	......	......	......	......	2 286	20 550	Hitchcock, Sill. Jour. [2] xx, 255.
8. Elton Lake, European Russia,	1 2729	3.830	......	19.750	......	5.319	......	......	0 230	29 130	H. Rose, Rep. Onondaga Salt Springs, 1851.
9 Great Salt Lake, Utah,	1.170	20 196	Trace.	0 252	......	......	......	......	1 834	22 422	Gale, Stansburg's Exped. 419.
10. Syracuse, (Old wells,)	1.10499	13.239	0.083	0 046	0 569	......	0.014	0.002	......	13 953	Beck, Rep. Min., N. Y., 110.
11. " (New wells,)	1 1426	17 690	0.156	0.119	0.573	......	......	0.002	......	18 540	G. H. Cook, Rep. Onond. Salt Sp., 1851.
12. Salina, N. Y.,	1 1138	14.042	0.144	0 110	0 551	......	......	0.003	......	14.850	" " " " "
13. Liverpool, N. Y.,	1 1269	15 100	0.190	0 111	0 454	......	......	0 002	......	15.860	" " " " "
14. Geddes', N. Y.,	1 10601	13 066	0 203	0.079	0 493	......	0 010	0 004	......	13 855	Beck, Rep. Min., N. Y., 110.
15. Montezuma, (Old deep well,)	......	7 372	0.153	0 030	0.431	......	0 002	......	......	7 988	Chilton, Sill. Jour. vii, 344.
16. " (East side,)	1.07543	9.335	0 140	0.100	0 525	......	0.018	0.002	0 008	10 128	Beck, Loc. Cit.
17. " (well of 1859,)	......	15.061	0.370	0.030	0.278	......	......	......	......	15 739	Webb, Rep. E. Saginaw Salt Co., Feb. 1860.
18. Galen, N. Y.,	1 0544	8.424	0.049	0.029	0.207	......	0.009	......	0.004	8 712	Chilton, Sill. Jour. vii, 349.
19. McConnelsville, Ohio,	......	14 000	0 780	1 040	......	......	......	......	......	......	Mitchell, 1 Ohio Rep. 61.
20. Equality, Gallatin Co., Ill.,	1 047	6.24	0 40	0 21	0.34	......	......	......	......	7 200	Prof. G. H. Cook, communicated.
21. Kanawha, Va.,	1 073	7 309	1.526	0.374	......	......	......	......	......	9.209	Cook, Onondaga Rep., 1854.
22. Saltville, Va.,	1.198	25 975	......	.....	0 322	0 104	......	......	......	26 400	" " " "
23. Salt Lake, Texas,	1.187	24.173	......	......	0 204	0 522	......	......	......	24 899	" " " "
24. Tipaquera, Bogota, S. A.,	....	15.270	9 300	4.500	......	......	......	0 270	0 020	29 360	Cheyne, Sill. Jour. xxxii, 84.
25. Cheshire, Eng., (Brit. Salt Co.,)	1.205	25.322	0 065	0.065	0 494	......	......	0.052	......	26 000	Dr. Wm. Henry, Repert of Arts [2] xvii, 295.
26. Dieuze, France, (Springs,)	1.120	15.214	......	0.275	0 328	0 582	......	......	......	16 399	Cook, Onondaga Rep., 1853.
27. Scribner's, Grand Rapids,	1 01752	1.737	0.276	0.072	0 131	0.005	0.001	......	......	2 334	Prof. L. R. Fisk, from surface brine.
28. East Saginaw Co., (617 ft.,)	1 172	19.088	0.537	1.241	......	......	......	......	0.225	21.091	Webb, Rep. E. Saginaw Salt Co., Feb., 1860.
29. " " "	1 170	17.912	2 142	1.522	0.116	......	......	0 105	0 220	22 017	P of. S. H. Douglass, in Detroit Free Press.
30. " " " (649 ft.,)	1 177	16.871	3 287	1.774	0 098	......	0.050	0 012	0.065	22 157	Chilton, communicated by W. L. Webber, Esq.
31. Sec. 25, T. 15 N., 1 W., Midland Co.,	1 0132	1.616	0.093	0.125	0.076	......	0 015	......	......	1 926	Houghton, Report 1838, p. 26.
32. Secs. 2 and 11, Chesterfield, Macomb Co.,	1.0057	0.549	0.013	0.037	0.015	......	0.014	0.001	......	0.629	" " " p. 29.

Dr. Houghton in his report of 1838, gave the results of analyses of 20 different brine springs from our State, two of which have been reproduced in the table. Of these springs, three were situated upon the Tittabawassee river, in Midland county, seven near the Grand river, two near the source of navigation of Maple river, in Gratiot county, two near the Maple river, in Clinton county, and one near the Saline river, in Washtenaw county. The solid constituents of these brines contained from 58 to 87 per cent. of pure salt, the general range being 70 to 86 per cent. The purest brine was found on section 24, T. 15 N., 1 W., Midland county, on the the Tittabawasse river, half a mile above the mouth of Salt river.

The following table will also prove useful for general reference:

TABLE *Showing the number of bushels of Salt made at the Onondaga Salt Springs, New York, since June 20th,* 1797, *which is the date of the first leases of lots.*

DATE.	BUSHELS.	DATE.	BUSHELS.
1797	25 474	1829	1,291,280
1798	59,928	1830	1 435,446
1799	42 474	1831	1,514,037
1800	50,000	1832	1,652,985
1801	62,000	1833	1,838,646
1802	75,000	1834	1,943,252
1803	90,000	1835	2,209,867
1804	100,000	1836	1,912,858
1805	154 071	1837	2,167 287
1806	122,577	1838	2,57[illegible],033
1807	165,448	1839	2,864 718
1808	319,618	1840	2 622,305
1809	128,282	1841	3,340,769
1810	450 000	1842	2,291,903
1811	200,000	1843	3,127,500
1812	221,011	1844	4,003,554
1813	226,000	1845	3,762,358
1814	295,000	1846	3,838,851
1815	322,058	1847	3,951,355
1816	348 665	1848	4,737,126
1817	408,665	1849	5,083,369
1818	406,540	1850	4,268,919
1819	526,049	1851	4,614,117
1820	548,374	1852	4,922,533
1821	558,329	1853	5,404,524
1822	481,562	1854	5,803,347
1823	726,988	1855	6,082,885
1824	816,634	1856	5,966,810
1825	757 203	1857	4 312,126
1826	811,023	1858	7,033,219
1827	983,416	1859	6,894,272
1828	1,160,888	1860	5,593,447
Total,			130,737,157

The following is an approximate statement of the amount of salt manufactured in the United States during the year 1859:

	Bushels.
Massachusetts, (mostly in vats along the shore,).....	15,000
Onondaga salt works, N. Y..........................	6,894,000
Pennsylvania, (Alleghany and Kiskiminetas rivers,)	1,000,000
Virginia, (Kanawha and King's works,)...........	1,900,000
Kentucky, (Goose Creek,).........................	300,000
Ohio, (Muskingum and Hocking rivers,)...........	1,500,000
Ohio, (Pomeroy and West Columbia,).............	2,500,000
Illinois, ...	5,000
Texas, ...	20,000
Florida, ...	100,000
Total,..	14,234,000
Foreign salt imported into the U. S. for the year ending June 30th, 1857,..............................	17,165,000
Foreign and domestic salt,..........................	31,399,000
Export of domestic salt,......................576,000	
" foreign salt,.......................131,000	
	707,000
Annual consumption of salt in U. S.,..........	30,692,000

Which for each individual amounts to,................52½ lbs.
In Great Britain it is,...25 "
In France, ...15½ "

Receipts of salt at Detroit for two years:

1859, ...52,203 bbls.
1860,...58,212 "

Receipts and shipments of salt at Chicago for seven years:

	Receipts, bbls.	Ship'ts, bbls.
1852,....................................	92,907	59,338
1853,....................................	86,309	38,785
1854,....................................	176,526	91,534
1855,....................................	170,633	107,993
1856,....................................	184,834	82,601
1857,....................................	209,746	90,918
1858,....................................	333,988	191,279
1859,....................................	316,897	250,467
1860,....................................	223,018	164,409

Of the shipments for 1859 and 1860, the following amounts were returned to Michigan:

	1859.	1860.
By Mich. C. R. R., ...	4,507 bbls.	2,478 bbls.
By Mich. S. R. R., ...	5,253 "	2,260 "
Total, besides shipments by lake, ...	9,760 bbls.	4,738 bbls.

About one-third of the fine salt blocks at Onondaga, N. Y., are worked with coal, which is furnished from Pennsylvania at $3,00 per ton. The use of coal has reduced the price of hard wood at the works from five and six dollars per cord, to $3 50. The latter sum corresponds to $2 33 for "soft" wood, and $2 92 for "mixed" wood. The price of barrels is at present about 26 cents. The prime cost of a barrel of salt (280 lbs.) at Onondaga is stated to be 95½ cents. At Kanawha it is 87½ cents.

The solar salt manufacture was carried on at Onondaga in 1858, by 28 different parties, using an aggregate of 30,786 covers, and occupying 8,403,840 square feet, or nearly 193 acres of surface. In 1860 the whole number of covers has increased to 36,302, occupying more than 207 acres of surface.

The fine salt manufacture was carried on in 1858 by 104 separate parties, who used an aggregate of 312 blocks and 16,434 kettles. No additions have been made to the close of 1860.

The aggregate value of the solar works, at $40 the cover, is, ...	$1,452,080
That of the fine salt works at $4,000 the block, is, ..	1,240,000
Total capital in salt manufacture, ...	$2,692,080

About 21 per cent. of all the salt manufactured at Onondaga is solar salt. This, it will be seen, requires a larger outlay of capital than the 79 per cent. of fine salt. The cost of manufacture of the coarse salt is, however, less, so that while one of the elements of the prime cost of coarse salt is greater another is less than the corresponding one for fine salt.*

The total annual produce of salt in the United Kingdom, is

*For nearly all my information relative to the salt manufacture at Onondaga, I have depended upon the Annual Reports of the Superintendent, for which I am indebted to Supt. V. W. Smith, and Prof. Geo. H. Cook.

1,462,045 tons, which, at 2,000 lbs. per ton, amounts to 52,215,893 bushels. The total exports and their value for three years, are as follows:

1855,	630,154 tons, valued at		£268,857
1856,	745,513	" "	276,242
1857,	651,766	" "	239,969

The principal salt producing districts in England are Cheshire and Worcestershire. It is mostly manufactured from rock salt. At Northwich, in the former county, the bed of salt is not less than 60 feet thick, a mile long, and 1,300 yards broad.

Salt is extensively manufactured from sea-water on the shores of the Mediterranean, in the south France, and on the western coast. At the saline of Berre the evaporating surfaces cover an area of 815 English acres, and the annual manufacture is 20,000,000 kilograms (2,205 lbs. each), or 787,500 bushels. The saline of Baynas yields annually 20,000 tons (757,500 bushels), 1,550 tons sulphate of soda (Glauber's salt), worth 30 francs the ton, and 200 tons of chlorid potassium, worth 360 francs the ton.

The total manufacture of salt, in France, in 1847, was as follows:

	Tons.
Salt marshes of the Mediterranean,	263,000
Western coast,	231,000
Salt springs and a mine,	76,000
	570,000

This amount, reckoning 1,000 kilogrammes to the ton, is equal to 22,443,750 bushels, and gives occupation to 16,650 workmen.*

Sea water is extensively evaporated by the Biscayans, on the shores of Spain and Portugal. The salines of the lagoons of Venice cover an area of about 1,630 English acres. The salt mines of Central Europe have been celebrated for ages. Those of Vieliczka and Bochnia, in Galicia, are well known. They be-

*For much valuable information on the manufacture of salt, especially in France, see a report "On the extraction of salt from sea-water," by T. S. Hunt, in Canada Geological Report for 1855, republished in Silliman's Journal, Vol. XXV [2] 361, May, 1858. Also Report of Prof. Geo. H. Cook, in Superintendent's Report of Onondaga Salt Springs, transmitted to the Legislature in 1853.

long to the extensive saliferous tract lying along both sides of the Carpathians, and embracing the mines of Wallachia, Transylvania, Galicia, Upper Hungary, Upper Austria, Styria, Salzberg and the Tyrol.

The total amount of salt annually produced by three of the leading nations of the earth, is as follows:

Great Britain,	52,215,893	bushels.
France,	22,413,750	"
United States,	14,234,000	"
	88,893,643	"

Besides the use of salt for mechanical and agricultural purposes, it enters largely as an article of food into the consumption of all classes of people; and it seems, like water and many other natural products, to have been provided with special reference to the physiological constitution of man. It is equally sought by the lower animals, especially the Ruminantia and Pachydermata. Bees, even, are fond of sipping it from a state of solution. Mungo Park says* that in the interior of Africa "the greatest of all luxuries is salt. It would appear strange to a European to see a child suck a piece of rock salt as if it were sugar. This, however, I have frequently seen; although in the inland parts, the poorer class of inhabitants are so very rarely indulged with this precious article, that to say a man eats salt with his victuals, is the same as saying he is a rich man. I have myself suffered great inconvenience from the scarcity of this article. The long use of vegetable food creates so painful a longing for salt, that no words can sufficiently describe it." Burchell states† that he sometimes had to send 90 miles for a gallon of salt.

The consumption of this article for food increases in the direct ratio of the average refinement of a people, or of the world. We can therefore see no limit to the demand. This will continue to increase most rapidly in those regions where population and improvement are making most progress. In this respect, no part of the world will compare with the great

*Travels, Vol. I., p. 280. †Travels in S. Africa.

Northwest. When, in addition, it is remembered that salt has long been used in some countries as an improver of the soil, and that recent researches* have shown it to be well adapted for this purpose, there is no reason to fear that the manufacture can ever be overdone. There are no evidences that the rapidly increasing supply of Onondaga salt has perceptibly affected the price for the period of 40 years.

Such being the facts, the vast geographical extent of the salt basin of Michigan, together with the extraordinary strength of the brine, furnish strong reasons to anticipate that at no distant day Michigan will be the leading salt-producing State of the Union; and a judicious public policy will be shaped with reference to forwarding this result.

PEAT, LIGNITE AND OTHER BITUMINOUS DEPOSITES.

Allusion has been made in a former part of this report, to the existence of numerous deposites of Peat, scattered over the surface of the Lower Peninsula. This substance is composed almost entirely of vegetable matter, which is the distinguishing characteristic of the luxuriant soils of the "prairie" States. Properly commingled, therefore, with our warm gravelly soils, the result would be a union of the exceliencies of two soils quite distinct from each other. Impressed with a vague idea of the agricultural value of peat, the farmer has not unfrequently strewn it in a crude state upon his fields and been disappointed at the temporarily injurious effects produced. It must be remembered, however, that peat is vegetable matter in a tate of *partial* decomposition; and if it were not actually injurious in this state, it could be of no use, as plants assimilate only *inorganic* or *disorganized* matter. But partially decomposed vegetable matter is made up to a great extent of various vegetable acids which impart a *sourness* to the soil, and prove a positive injury to crops. Obviously, therefore, the decomposition of the peat must be *completed* before it is suitable as an application to the soil. Various means are recommended for

* Yale Agricultural Lectures, p. 181.

this purpose by writers on scientific agriculture, but as it is not my intention here to enlarge upon this subject, I only allude to two.

First of all, the peat or muck should be thrown out and left where it can be exposed to the process of alternate soaking and drying, and if possible also to the action of frost.

Secondly, it may be mixed with lime, which, as an alkaline agent, will neutralize the acidity, and at the same time facilitate decomposition. When thus mixed, it is much more promptly prepared for use. The lime for this purpose has not to be quarried from a distant ledge and burned in a kiln. Nature has placed it in the form of marl, in immediate juxtaposition with the peat which needs its agency. Indeed the farmer can in many cases load his cart with the mixed deposits without even moving his team from their tracks. I hardly know a more striking adaptation of natural means for the accomplishment of a necessary object. The porous nature of our soils suffers their soluble constituents to be carried away to the lower levels, where peat and marl are accumulating, and where the growths of ages unknown, have been adding a thousand fold to the nutritive elements brought down from the soils of the contiguous hill slopes. These depositories of agricultural force, a good economy will not fail to appreciate and apply to the recuperation of declining wheat lands.

While, however, the application of peat as a fertilizer to the soil is its most obvious use in a purely agricultural region, it cannot be said that this is its principal, or even its most important application. Though in a country like our own, covered with primitive forests, the value of peat as a fuel is almost unknown, the amount consumed in older countries is truly enormous. The bogs of Ireland are estimated to occupy 2,830,000 acres. Two million acres, at an average depth of nine feet, assuming peat to be but one-sixth the value of coal, will furnish an amount of fuel equal to 470,000,000 tons of coal, worth thirteen hundred millions of dollars. For the purposes of ordinary fuel, the raw peat is prepared by subjecting it between cloths, to the pressure of a powerful hydraulic press. This condenses it to one-third of its original volume, and three-fifths of its original weight, through loss of moisture. At the large peat bog near Liancourt, on the Northern Railway, nineteen leagues north of Paris, the peat after having been thoroughly mixed and worked together, is moulded under great pressure into small bricks, which, when dried, are heavier than water. The moulded peat is worth in Paris 20 francs the ton of 1,000 kilogrammes, (2,204 pounds avoirdupois.) The amount raised at this bog annually is 10,000 to 12,000 tons. At Rheims 14,000 tons are annually produced. A peat bog in the vicinity of New York city, six feet deep and forty acres in extent, is stated by Prof. Mather to have yielded a fuel which retailed for $4 50 per cord, realizing $4,500 per acre, a little more than a third of which was expenses.

For mechanical, and not unusually for domestic purposes, the dried peat is first converted into a coke or charcoal, of which it yields from 40 to 42 per cent. Peat charcoal sells in Paris for about the same price as wood charcoal, or 13 francs the 100 kilogrammes—the relative prices of wood or peat charcoal, mineral coal and wood, being as the numbers 13, 4½, 4¾ respectively. This proportion would of course vary with the relative abundance of peat, wood and mineral coal, in any country. Peat coke occupies about the same space, weight for weight, as ordinary coke, and only half that of charcoal, having a specific gravity of 1.040, that of charcoal from hard woods averaging 0.505. For heating purposes, 7 tons of peat coke are equivalent to 6 tons of good coal coke. For the manufacture and working of iron, peat coke is pronounced decidedly superior to charcoal, both in consequence of its greater heating property and its production of a superior quality of iron. It is extensively employed in preference to any other fuel in many of the furnaces of France, Bavaria, Wurtemberg, Bohemia and Sweden. For steam producing purposes, compressed peat has been proved at least equal to any other fuel. A factitious coal is prepared from peat by the Dublin Steam Navigation Company, 10 cwts. of which generate the same steam power as 17½ cwts. of pit coal. Peat is very extensively employed on the steamers which ply in the waters in and about Ireland, and even upon the river Shannon, in the midst of a coal bearing country. Some of the prepared peats of France are also said to be economically employed for stationary steam engines, and even for locomotives.

The uses to which peat has been profitably applied do not stop even here. A company exists at Kilberny, in Ireland, having a factory in operation in which they produce from peat, Tar, Paraffine, Oil, Naphtha, Sulphate of Ammonia, and a Gas, the combustion of which is applied to the manufacture of Iron. The most thorough and extensive manufacture of these products, however, seems to be effected by Messrs. Babonneau & Co., at Paris. According to Mr. Armand, the skillful chemist of this establishment, good peat yields, on an average, about 40 per cent. of charcoal, 15 to 18 per cent. of crude oil containing paraffine, 36 per cent. of water containing carbonate, acetate and sulphydrate of ammonia, and a little wood spirit, besides 7½ per cent. of inflammable gases and loss. The ammonia is equal to 2 per cent. of *sal ammoniac*. The oil, by distillation, is separated into a light oil or *naphtha* which is burned for illumination, in lamps of a peculiar construction, and a heavy, less volatile portion which is used for *lubricating* machinery, or is mingled with fat oils for burning in ordinary lamps. There is obtained besides, a portion of solid *bitumen or pitch* amounting to 4 or 5 per cent. of the dried peat. The *paraffine*, which is dissolved in the oils, is separated by exposing them to cold, and is afterwards purified. The yield of this product is 2 or 3 per cent. of the peat. When pure, it is a white, fusible crystalline solid, devoid of taste or smell, much resembling spermaceti in appearance, and like it employed in the manufacture of candles. The price of paraffine in France is a little more than one franc per pound.

The gas evolved during the distillation of peat may be employed, as at Kilberry, in Ireland, for the purposes of *heating*, or it may be mixed with the gas obtained by the decomposition, at a high temperature, of the crude oil from peat. In this way an *illuminating gas* is obtained which has three and four-tenths times the illuminating power of coal gas, while the yield is equal to that from coal.

The solid bitumen resulting from the distillation of peat may be employed like asphalt in the preparation of *mastic for paving.* Even the crude peat, by being mixed after drying with 10 to 15 per cent. of coal tar, and boiled for several hours, dissolves into a viscid liquid, which, when cooled, is solid, and resembles asphalt. The crude residues from the rectification of the oil of peat are burned in proper apparatus, and furnish abundance of *lampblack.*

For the production of *gunpowder*, many varieties of peat are superior to the charcoal of dogwood and alder.*

The reader, perhaps, will hardly deem it credible that so great a variety of commercial products is obtained from a substance so common and so little valued as the "muck" with which our "swamps" are filled. As all such doubts arise from ignorance of the properties of peat, I present below a convenient synopsis of the products and uses of this substance:

1. Crude peat as a fertilizer for the soil.
2. Prepared peat and peat-coke as fuel.
   (*a*) For domestic and ordinary heating purposes.
   (*b*) For the generation of steam.
   (*c*) For the manufacture and working of metals.
3. Peat for the manufacture of gunpowder.
4. Peat or bitumen from peat for paving purposes.
5. Crude oil for purposes of lubrication, illumination and gas-making.
6. Petroleum for burning in lamps.
7. Paraffine for the manufacture of candles.
8. Light, inflammable *gas* for heating.
9. Illuminating gas of superior quality.
10. Lampblack.

The value of peat for any or all of the above purposes will obviously depend upon its freedom from earthly deposites. In

* For valuable information on the subject of Peat, the readeri s referred to "Taylor's Statistics of Coal," and T. S. Hunt's Chemical Reports, in the Canada Geological Reports for 1850 and 1855.

those cases where a bog has grown with the growth and decay of Sphagnum, or other bog mosses, the peat is often composed of almost pure vegetable matter. In other cases, where the bog has been periodically inundated, as around the margins of some lakes and ponds, more or less of earthly sediment will be found mixed with the peaty materials. A large proportion of our principal peat bogs, however, will compare favorably in purity with those in foreign countries, to which I have already alluded.

It will of course be inferred that the bed of lignite which I have described as occurring on the shore of Grand Traverse Bay, possesses all the capabilities of ordinary peat. Should the spontaneous flow of petroleum from the rocks ever be materially diminished, the same product may be very cheaply distilled, as is done in foreign countries, from lignite and peat as well as from coal.

Although it might be better to speak of Rock Oils or Petroleum under a distinct head, still the subject is here naturally introduced, and I proceed to append the few remarks which I have to offer on this subject.

The distillation of bituminous shales and mineral bitumens is carried on to a great extent both in England and on the continent. To this class of matters belong the so-called Boghead and cannel coals, as well as the bituminous minerals of various parts of France and Switzerland. Here belongs the black bituminous shale of Canada West, and Thunder and Grand Traverse Bays in this State, which will undoubtedly prove uncommonly rich in bituminous matter. Indeed, the abundant spontaneous distillation from shales of the same age, which has supplied the oil wells of Pennsylvania and Ohio, is an evidence that the products of artificial distillation would prove correspondingly rich. These substances yield, in general, the same products as peat. The amount of paraffine, however, is said to be less, and the residue left from distillation is, unlike that from peat, comparatively worthless. A yield of five per cent of bituminous matter qualifies the shales in France to be economically worked. The yield of our shales has never been accurately ascertained, and I have no means at my disposal for the determination of these important questions.

Shales thus bituminized have an existence in our State, about which there can be no question. I have elsewhere expressed the hope that they will yet be found to yield a spontaneous flow of Petroleum, like those of neighboring districts. The belt of country along which experiments might be made extends from Wayne county to Port Huron, and from Thunder to Grand Traverse Bay. The geological relations and the surface indications are such, especially along the southern belt, that a few borings would be fully justified. A few years ago, as I am informed by Mr. F. P. Bouteller, a boring for water was undertaken beneath a saw mill in the township of Greenfield; Wayne county. After the drill has passed through a bed of bluish shale at the depth of 70 or 80 feet, it was suddenly wrested from the hands of the workmen by the violent escape of a fetid gas which threw up water and sand to the height of several feet. By accident, the stream of gas was ignited, and sent a column of flame to the roof of the mill, which had to be removed. All efforts to entinguish it proved futile for several hours, when the furnace pipe was placed over the well to guide the flame. This, to the great relief of the owner, had the effect of smothering the fire. Grateful for his escape, he effectually closed the door against any further eruption of the nether fires by promptly filling the hole with stones well rammed down; and has stoutly persisted in refusing to allow any further experiments of this dangerous character upon his premises. Similar phenomena have been witnessed at various points along the shores of the St. Clair river and lake.

Inflammable gas is the product of the distillation of petroleum, and it is not improbable that by extending explorations below the horizon of the gas, the reservoir of oil would be reached.

WELLS AND SPRINGS.

The late successful boring of several artesian wells in the

southern part of the State, has created a very general desire to know to what extent artesian borings would prove successful in other parts of the State. Several unsuccessful borings have been made at points where the work has been directed rather by empiricism than by any adequate knowledge of the existence of such a geological structure as could furnish reasonable grounds for the expectation of success.

From what has already been stated of the general conformation of the strata underlying the Lower Peninsula, the accumulation and retention of vast reservoirs of water in these great peninsular dishes, will appear obvious and necessary. Rains falling upon the surface percolate downwards until the water reaches an impervious stratum along which it flows till it reaches the lowest depression of that stratum, somewhere beneath the center of the State, and some hundreds of feet from the surface. The water-bearing strata are, therefore, porous sandstone, immediately underlain and overlain by impervious strata of an argillacious or calcareous character. Each porous sandstone stratum thus underlain and overlain throughout our whole series, becomes in this manner surcharged with water admitted at its outcrop. It is obvious, now, that by boring down at any point within the circuit of the outcrop of a water-bearing stratum, until that stratum is pierced, the water will rise through the hole to a point on a level with the rim of the basin which holds the water. If the place of boring is lower than that point, the water will rise to the surface and overflow; if higher, it will not.

In consequence of the general rise of the surface of the peninsula from the lake shores toward the interior, the outcrops of the strata occur, as a general rule, at lower levels than the points within the basins which they form; and artesian wells cannot be a thing of general occurrence. In the southern part of Jackson, and the northern part of Hillsdale counties, however, the sandstones of the Napoleon and Marshall Groups outcrop at levels considerably higher than the general elevation of the peninsula, and it is likely that the impediments to a free circulation of the water, in these strata, prevent it from sinking, in these elevated sections, to the level of the lowest portions of the basin in remote parts of the State. As a consequence, artesian borings *might* prove successful throughout the southern half of Jackson county, and the eastern portion of Calhoun, if continued down to the bottom of either of these groups.

It must not be supposed, however, that the artesian wells of Jackson are supplied from this source. If I have succeeded in the identification of the rocks in that vicinity, these wells are supplied from the Parma Sandstone Albion is outside of the rim of this formation, and the wells there have to be continued down to the bottom of the Napoleon Sandstone. Marshall is outside of the rim of this, and rests just upon the rim of the outcropping Marshall Group; and hence I should not expect that the contained waters would rise to the surface. The artesian (salt) wells of Grand Rapids are supplied from the Napoleon Group, the water being salted from the group immediately above. The wells at Saginaw issue from the same sandstones, and are salted in the same way. In the southern part of Jackson and northern part of Hillsdale counties, where the streams have cut through these rocks, the contained waters rush out in extended chains of most beautiful and copious springs of pure water. Adrian is located upon the argillaceous strata of the Huron Group, and the first water-bearing stratum which would be reached is included in the Monroe limestones, perhaps 250 feet below. But the surface slopes gradually toward Lake Erie, so that the hydrostatic pressure would not be adequate to an artesian overflow. Ann Arbor is supposed to lie within the rim of the Marshall and Napoleon sandstones, but the considerable elevation of this place precludes all expectation of an overflow. The artesian wells at Toledo do not reach the solid rock at all, though this has been unsuccessfully explored to a considerable depth. The alluvial deposites, which are here of great depth, are made up of alternating sandy and argillaceous beds, which slope gradually toward the bed of the lake, and of course outcrop successively on the higher levels, several miles back from the lake shore. These, like the more solid water-bearing strata, carry the water from the surface along impervious floors, until it passes under the city, and finds its way into the artesian borings.

From what has been said of the occurrence of outlying patches, or small detached basins of carboniferous rocks, and the gently undulating character of the whole system, it will at once be inferred that besides the great basins just alluded to as reservoirs of water, there must be numerous smaller local basins. The indications seem to justify the conclusion that the wells at Jackson are supplied from a local basin. It appears, therefore, that a reliable opinion on the prospect of success at any particular point involves not only a knowledge of the general conformation of the rocks, but also an acquaintance with the special geology of the region in question.

In those portions of Calhoun, Jackson and Hillsdale counties which are situated over the outcrops of the Napoleon and Marshall sandstones, very many of the common wells terminate in these rocks, and from them derive their supply of water. Nearly all the wells of the Lower Peninsula, however, derive their supply from the sands of the Drift. The materials of the upper portion of this formation have been, by geological agencies, considerably assorted, so that beds of arenaceous materials alternate with beds of argillaceous materials, as in the underlying rocks. There is, however, no general stratification of these deposites Every bed of sand is comparatively local. No general parallelism can be traced among them. The argillaceous layers of the drift may be compared to a pile of wooden bowls thrown confusedly together—the interspaces being filled with sand. At one point, a well will be found to be within the rim of a given bowl, while at a very short distance from that, an excavation would prove to be outside of the same basin and would have to be carried perhaps to a much greater depth before reaching the bottom of the basin which underlies. On the University grounds, wells are sunk 70 to 80 feet before reaching water, while at the Observatory, which is 42 feet higher, the water rises within six feet of the surface. The latter well is obviously supplied from a local basin which occupies a higher level.

The purity and salubrity of well and spring water, in the Lower Peninsula of Michigan, are generally very great. An analysis of the water from the well on the north side of the University campus, was made by T. C. McNeill, A. B., of the Laboratory of Applied Chemistry, with the following result:

Depth of well, 70 feet 8 inches.
Temperature of water, 50° Fah.
Free carbonic acid in 100 parts, .015598.

Solid constituents:

Carbonate of lime,..............................	0.017800
Carbonate of magnesia,..........................	0.006058
Carbonate of iron,..............................	0.000290
Chlorid of sodium,..............................	0.000448
Sulphate of soda,...............................	0.000507
Carbonate of soda,..............................	0.000152
Sulphate of potash,.............................	0.000678
Silicic acid,...................................	0.000730
Organic matter,.................................	0.002300
Total,..	0.028963

The wells of Detroit, and much of the region along the lake and river shore, from Toledo to St. Clair, are sunk in lacustrine deposites, which impart a greater per centage of organic and soluble matter. The following analysis was made by Prof. S. H. Douglass, in 1854, for the Board of Water Commissioners of the city of Detroit. The water was taken from a well at the residence of Amos T. Hall, on Woodward Avenue:

Chlorid of potassium,...........................	0.011000
Chlorid of sodium,..............................	0.072520
Chlorid of magnesium,...........................	0.034760
Sulphate of potassa,............................	0.010450
Sulphate of lime,...............................	0.028260
Silica,...	0.002370
Carbonate of lime,..............................	0.039190
Carbonate of iron,..............................	0.001020
Total,..	0.199570

The water of Detroit river at the same time contained the following constituents:

Sulphate of potassa,	0.000283
Sulphate of soda,	0.000750
Phosphate of lime,	0.003110
Alumina,	0.001050
Silica,	0.000500
Carbonate of lime,	0.003300
Carbonate of iron,	0.000814
Total,	0.009807

By far the most important mineral waters of the Lower Peninsula are those charged with chlorid of sodium. The ferruginous sandstones of the lower part of the State, give origin, however, to numerous springs which are strongly chalybeate, while the bituminous rocks of the Huron and Upper Helderberg groups, become the source of strongly sulphureous waters. No formal investigations have been made of any of these springs. The following analysis, however, by Mr. McNeill, before quoted, was made upon the water of a spring issuing upon the land of Solomon Mann, Esq., Ann Arbor:

Temperature, 50° Fah.
Specific gravity, 1.001.

Constituents of the solid matter:

Carbonate of lime,	0.022800
Carbonate of magnesia,	0.008936
Carbonate of iron,	0.000468
Chloride of sodium,	0.000488
Iodide of sodium,	trace.
Sulphate of soda,	0.000971
Carbonate of soda,	0.000042
Sulphate of potash,	0.000531
Silicic acid,	0.001200
Organic matter,	0.002500
Total,	0.037936
Free and partially combined carbonic acid,	.028500

The quantity of iron in this water is greater than that in the chalybeate waters of Bath, England, and Karlsbad and Teplitz, in Bohemia, though the total solid constituents are considerably less.

The sulphur springs of the southern portion of the State are exceedingly numerous, and I shall take the space in the present report to allude particularly to only two.

A very remarkable spring occurs on section 22 (?), in the township of Erie, Monroe county. It is situated within the marsh which borders the lake, about one mile from the lake-shore and four miles south east from Vienna. The spring has to be reached by boat. It is found occupying a conical depression, about 200 feet in diameter and 45 feet deep. Some time before reaching the spot the sulphureous odor can be detected, when the wind is favorable. At the distance of 30 rods the water of the bayou has a sulphuretted taste, and a whitish deposite can be seen on the stems of aquatic vegetation. At the time of my visit the rim of the basin was 18 inches under water, but later in the season the water subsides, and the rim is converted into a fine walk around the pool. Under these circumstances the flow of water from the spring forms a stream 10 feet wide and 3 feet deep, with a considerable current.

Another interesting locality is found on the south side of the Raisin river, nearly opposite the Raisinville lime quarries, in Monroe county. Here is a chain of sulphur springs on the land of Robert Talford. On approaching the locality sulphuretted fumes are very distinctly perceived. The water boils up in very copious quantities at more than half a dozen points within the area of a quarter of an acre. A copious, white—almost snow white—deposite lines the banks and bed of the stream which flows off from these springs. The several rills uniting form a stream capable of turning a small mill, or perhaps discharging 1200 gallons of water per minute. Through a log erected in one of the springs, the water rises 8 or 10 feet. In the midst of the group is a fine spring of sweet water.

The evidences of sulphur here are equal to those seen at some of the most celebrated watering places. It is a cause of astonishment that efforts have not long since been made to render this a place of resort for invalids and others. The springs are located in a dry, elevated limestone region. The surroundings, though not picturesque, are diversified and agreeable. The water is strong and copious. Access is comparatively easy by public conveyance on the plank road 8½ miles from Monroe.

Three other groups of springs of equal copiousness exist in the immediate neighborhood, and numerous others are scattered throughout the county.

CHAPTER VII.

PHYSICAL GEOGRAPHY, TOPOGRAPHY, HYDROGRAPHY, METEOROLOGY.

It becomes necessary in the present hasty report, to omit all reference to these subjects except what is embraced in the following table, which is partially reproduced from Higgins' Report, as Topographer to the Geological Survey, (Rep. 1839, p. 61,) partly from Foster and Whitney's Report on the Lake Superior Land District, Part I, pp. 18, 38 et seq, and is otherwise compiled from original observations, and other unpublished data:

ALTITUDES of various points within the State of Michigan.

LOCALITIES.	Above Lake Huron.	Above the Sea.
Lake Erie,	......	565
Detroit River at Detroit,	......	568
Base of old Capitol at Detroit,	18	596
Wayne Station,*	80	658
West line, Wayne county,	138	716
Ypsilanti Station,	135	713
Geddes' Station,	168	746
Ann Arbor Station,	193	771
University buildings, Ann Arbor,† (by level from depot,)	298	876
Observatory, Ann Arbor, (by level from depot,)	340	918
Delhi, Washtenaw Co.,	239	817
Scio, " "	250	828
Dexter, " "	281	859
Chelsea, " "	338	916
West line Washtenaw Co., on railroad,	437	1015
Francisco Station, Jackson Co.,	440	1024
Grass Lake " " "	411	989
Leoni, " " "	401	979

* Heights of points along M. C. R. R., unless otherwise designated, have been communicated by Thos. Frazer, Esq., of the Central Office, Detroit.

† The corrected mean of the Barometer at the University, for 9 mos., ending Feb., 1855, was 29.047 inches, which corresponds to an altitude of 809 feet, while the height of the place of observation was supposed to be 891 feet.

LOCALITIES.	Above Lake Huron.	Above the Sea.
Michigan Center, Jackson Co., ...	363	941
Jackson, "	400	978
Barry, "	362	940
Albion, Calhoun Co., ...	365	943
Kalamazoo river, Albion, Calhoun Co., (Higgins,)	351	929
Half Way House, Wayne Co., "	54	632
Head of Spectacle Lake, Calhoun Co., "	373	951
Rice Creek, near Marshall, " " "	280	858
Honey Creek, Washtenaw Co., "	26[illegible]	844
Huron river, Ypsilanti, " "	100	678
Huron river, Dexter, " "	232	810
Sandstone Creek, Jackson Co., "	347	925
Outlet of Gillett's Lake, " "	354	932
Outlet of Grass Lake, " "	377	955
W. end of Prairie Ronde, Kalamazoo Co., "	278	856
Kalamazoo R., sec. 35, Augusta, " "	187	765
" Kalamazoo, " "	154	732
Crossing M. S. & N. I. R. R., at St. Joseph R., St Joseph Co., (Higgins,)	138	716
Branch, St Joseph R., sec. 35, Mattison, Branch Co., (Higgins,)	187	765
Bank of L. Michigan, New Buffalo, Berrien Co., (Higgins,)	100	678
Bank of Galien R., 10 miles E. of New Buffalo, Berrien Co, (Higgins,)	74	652
St. Joseph R., at Bertrand, Ber'n Co., (Higgins,)	53	631
Paw Paw R., Lafayette village, Van Buren Co., (Higgins,)	106	684
Bush Creek, near Mason, Van B'rn Co., (Higgins,)	76	654
Stony Creek, "crossing Northern R. R.," Ionia Co., (Higgins,)	82	660
Mouth of Maple River, Ionia Co., (Higgins,)	56	634
N. branch Raisin R., Lenawee Co., "	276	854
Hasler's Creek, Lapeer Co., "	265	843
Head of Belle R., " " "	414	992
Flint River, Lapeer, Lapeer Co., "	238	816
Shiawassee R., Owosso, Shiawassee Co., "	145	723
Village of Newberry, St Clair Co., "	284	862
Head of Mill Creek, " " "	368	946
Pontiac, Oakland Co., ...	336	914
Bass R. crossing, "Northern R. R.," Ottawa Co., (Higgins,)	56	634
Crossing Southern R. R., 4 m. W. of Monroe, (Higgins,)	51	629

LOCALITIES.	Above Lake Huron.	Above the Sea.
Cass R., T. 11 N., 5 E., 1 ft. above Saginaw R. at East Saginaw, (M. B. Hess,) supposing the river falls 1 ft. from there to the Bay,......	2	580
Birch Run, T. 10 N., 5 E., (M. B. Hess,)........	26	604
Pine Run, T. 9 N., 6 E., " "	105	683
Summit bet. Flint and Pine rivers, in T. 9 N., 7 E., (M. B. Hess,)......................	227	805
Flint river, T. 7 N., 7 E., (M. B. Hess,)........	120	698
Detroit Station, Detroit & Mil. R.,*............	......	575
Royal Oak, Oakland Co.,....................	79	657
Birmingham, " "	190	768
Pontiac, " "	349	927
Drayton Plains, Oakland Co.,.................	381	959
Waterford, " "	404	982
Clarkston, " "	415	993
"Clarkston Cut," " "	440	1018
Springfield, " "	438	1016
Davisburgh, " "	370	948
Holly, " "	340	918
Fentonville, Genesee Co.,....................	330	908
Linden, " "	291	869
Gaines, " "	267	845
Vernon, Shiawassee Co.,....................	181	759
Corunna, " "	185	763
Owosso, " "	154	732
Ovid, Clinton Co.,.........................	146	724
St. Johns, "	177	755
Dallas, "	157	735
Pewamo, Ionia Co., ...	153	731
Muir, "	67	645
Ionia, "	60	638
Saranac, (Boston,) Ionia Co.,..................	50	628
Lowell, Kent Co.,.........................	48	626
Ada, " "	75	653
Lamphier's Creek, (crossing, only,) Kent Co.,...	200	778
Grand Rapids, Kent Co.,.....................	54	632
Berlin, Ottawa Co.,.......................	91	669
Coopersville, "	54	632
Nunica, "	45	623
Mill Point, "	4	582
Grand Haven, "	4	582
Mean height of Lower Peninsula, (Higgins,)...	160	738

* For heights of points along the D. & M. R., I am indebted to Superintendent W. K. Muir.

LOCALITIES.	Above Lake Huron.	Above the Sea.
Lakes Huron and Michigan, (Higgins,)........	......	578
Lake Ontario, "	......	235
Sliding bank, entrance to Hammond's Bay, Lake Huron, (Halt 676,)......................	77	655
Bluff at Marble Quarry, E. end Drummond's I,..	98	676
Fort Mackinac, (Higgins,)..................	150	728
Old Fort Holmes, Mackinac I., (Higgins,)......	219	797
" " by Barometer, July, 1860,...	307	885
" " by Geological level, July, 1860,	318	897
" " according to Foster & Whitney,	315	893
Robinson's Folly, (Higgins,).................	128	706
" " by Geol. level, July, 1860,...	127	705
Bluff facing Round I., " " " ...	147	725
Summit of Sugar Loaf, " " " ...	284	862
Chimney Rock, " " " ...	131	709
Lover's Leap, " " " ...	145	723
Top of arch at Arched Rock, by Geol. level, July, 1860, ...	140	718
Top of arch at Arched Rock, by Barometer,....	138	716
To highest summit of Arched Rock, by level,...	149	727
Top of Buttress facing the lake at do. " ...	105	683
Principal Plateau of Mackinac Island, " ...	150	728
Upper Plateau of " " " ...	294	872
Summit of St. Joseph I., (T. N. Molesworth,)...	400	978
Lake Superior, (Foster and Whitney,).........	......	627

LOCALITIES.	Above Lake Superior.	Above the Sea.
Pie Island, N. shore L. Sup., (Foster & Whitney,)	760	1387
McKay's Mountain, "	1000	1627
Thunder Cape, "	1350	1977
St. Ignace, (estimated,) "	1300	1927
Les Petits Ecrits, "	850	1477
Pic Island, "	760	1387
Michipicoten Island, "	800	1427
Gros Cap, (estimated,) "	700	1327
Highest Point Porcupine Mts., "	1380	2007
Mt. Houghton, near head of Keweenaw Point, (Foster & Whitney,)....................	884	1511
Grand Sable, L. S., (transported materials,)....	345	972
Pt. Iroquois, " " "	350	977

PART II.

ZOÖLOGY.

CHAPTER XVIII.

REPORT OF THE STATE ZOÖLOGIST.

LANSING, Mich., Dec. 20th, 1860.

To PROF. A. WINCHELL, *State Geologist:*

SIR—I have the honor to transmit the following report of progress made in the Zoological department of the Natural History Survey of the State, during the past two years.

Owing to the limited appropriation made by the Legislature for the purpose of recommencing the Geological Survey of the State, and the desirableness of prosecuting the explorations in the Geological department with as effective a force as possible, I have been unable, as you are well aware, to devote but a part of my time to investigations in the department assigned me in the organization of the Geological corps.

The Zoological collections already made comprise such specimens as could be readily obtained without the sacrifice of much time, or detracting materially from the efficient progress of the Geological reconnoissance.

Very much remains to be done before an approximation to a complete knowledge of our fauna can be obtained.

From the nature of the subjects of investigation in this department, it is difficult, in a limited space, to give a satisfactory account of the exact progress of the work, or even to embody the results accomplished when so much remains unfinished.

The subjoined catalogue of the species known to inhabit our State, will, perhaps, best present an outline of the labor already performed, and at the same time furnish desirable information in regard to the geographical range of species.

In addition to the list here presented there are large numbers of specimens that remain to be identified and described, which will materially increase the number of known species in the State.

The fishes, insects, and crustaceans have not been worked up and for that reason have been omitted from the catalogue.

It may not be out of place in this connection to make a brief statement of the aims to be kept in view, and the results which may be expected to follow from the earnest prosecution of the study of the Zoology of our State.

From the intimate and important relations existing between man and the various branches of the Animal Kingdom, he is particularly interested in becoming acquainted with the forms, structure, metamorphoses, habits, and dispositions of the animate beings which surround him. He would thus be better fitted to act intelligently in availing himself of the benefits to be derived from those species that are capable of improvement by domestication, and at the same time be enabled to successfully maintain that influence and control over the economy of inferior organizations which his superior physical and mental developement, as well as interests, require of him. Dependent upon the animal kingdom, as he is to a great extent, for many of the comforts and luxuries of life, it would seem that the importance of a thorough investigation of the laws which govern this magnificent creation of living beings, and their relations to man's well-being and interests cannot be too highly estimated.

In the present advanced state of the abstract sciences, every branch of inquiry or investigation, no matter how trivial or unimportant it may in itself appear, tends directly to develop results that are of practical application in the varied pursuits of life.

Many illustrations of this fact might be adduced, and will undoubtedly present themselves to the minds of those who are familiar with the history of the useful arts. I will, however, cite but a single instance.

When it became known that sealing-wax, amber, and other resinous bodies, on being rubbed would attract pith-balls and other light substances, the discovery was looked upon as unimportant and trifling, and no one thought the knowledge capable of being made available for any practical purpose; yet from this small beginning the science of electricity has been developed, which, in its practical applications in the arts, no one in the present age would venture to set a limit. From the application of the principles of this science we are indebted for the increased facilities in the art of printing, by the process of electrotyping, improvements in the art of gilding, as well as for that wonder of the age, the magnetic telegraph, that brings by its network of wires the most remote places into almost instantaneous communication.

The so-called trifling experiments of philosophers, considered by many as beneath the attention of intelligent beings, have brought forth fruit abundantly, the influence of which on the world's progress can hardly be estimated.

Thus, in every department of knowledge, practical results are constantly presenting themselves as the inevitable consequence of progress in the purely abstract investigations of science.

An accurate scientific knowledge of the appearance, food, development, and mode of existence, of the various animal forms we are brought in contact with in our every day pursuits, as well as their varied relations to the vegetable and inorganic kingdoms of nature, is indispensable if we would derive practical benefit from the different classes of the animal kingdom and render them subservient to our prosperity and happiness.

Such knowledge to the agriculturalist would indeed be found of incalculable advantage; it would enable him to protect, as far as possible, the many species that confer direct benefits by furnishing various useful products, and to encourage the development of those that assist in protecting his crops, by preying on noxious forms, and thus preventing their inordinate increase, and at the same time he would be better prepared to adopt suitable measures for destroying and keeping in check those that by their depredations on his forest trees and grasses become most formidable enemies and the pests of civilization.

The army of weevils, Hessian flies, midges, chinch bugs, and cut worms attacking his wheat and other cereals, the numerous species of borers, curculios, locusts, and moths so destructive to his fruit and forest trees, all point to his interest in becoming better acquainted with the economy of nature, and studying more closely the varied phenomena presented by organic beings.

The intimate relations of Zoology to the other departments of science, might be cited as an incentive to a more general dissemination and increase of the knowledge of organic beings.

Geology derives important aid in its investigations from the application of the principles of Zoology; indeed, the rapid progress of the science of Geology at the present time is owing to the accurate investigations of the relations existing between the organic forms at present inhabiting our globe, and those fossil remains that are the index of the faunas and floras of past ages.

It is in fact in consequence of the aid furnished by the kindred sciences of Botany and Zoology that modern Geology has attained her proudest achievements.

In the State of New York alone hundreds of thousand of dollars have been expended in explorations for coal, when an examination of a few shells that abound in her rocks, would have shown that the entire geological formations of the State were below the coal bearing series of rocks, and that explorations for that mineral would consequently be fruitless.

In an educational point of view, a systematic knowledge of the animals inhabiting our State, their habits and relations to man and the surrounding world, would furnish a fund of materials for reflection and study, which, as a means of mental culture and developement, is capable of attaining a high rank among the studies considered essential in our institutions of learning, to a successful training of the intellectual powers.

At the present time, when a knowledge of the principles of

Natural History is considered indispensable to a finished education, the want of Museums in our State where the materials for the prosecution of this most interesting branch of study may be accessible to every one, is severely felt.

To supply this want, complete collections of the plants and animals of our State should be made so as fully to illustrate their systematic relations and affinities of structure, due prominence being given in their arrangement to the exhibition in a suitable manner of those species that are of benefit to the agriculturalist, as well as those that from their habits are continually warring against his interests by committing ravages that it is an object to keep within due bounds.

Aside from all this, the study of nature has a still higher significance than can be measured by any merely practical or pecuniary advantages accruing from its prosecution.

As the material expressions of the ideas of the Creator, the Supreme Intelligence of the Universe, the world of organic beings which He has created for man's contemplation and improvement, is certainly worthy the careful consideration of the highest faculties of the human mind.

In conclusion, I would make this public acknowledgment of my indebtedness to a number of scientific gentlemen for their disinterested assistance and encouragement, and to the public generally for the many acts of kindness shown to myself and party, during the progress of the survey.

M. MILES,
State Zoologist.

A CATALOGUE

OF THE

MAMMALS, BIRDS, REPTILES AND MOLLUSKS,

OF MICHIGAN,

BY M. MILES, M. D., STATE ZOÖLOGIST.

CLASS MAMMALIA.

UNGUICULATA.

ORDER CHEIROPTERA.

FAMILY VESPERTILIONIDAE.

1. Vespertilio Noveboracensis, *Linn*—New York Bat.
2. " fuscus, *P. de B.*
3. " subulatus, *Say.*—Brown Bat.
4. " phaiops, *Temm.*
5. " Caroli, "

ORDER RAPACIA.

SUB-ORDER INSECTIVORA.

FAMILY SORICIDAE.

Sub-Family Soricinæ.

6. Blarina talpoides, *Gray.*—Shrew.

FAMILY TALPIDAE.

7. Scalops aquaticus, *Fisch.*—Common Mole.
8. " argentatus, *Aud. & Bach.*—Silvery Mole.
*9. Condylura cristata, *Illiger.*—Star-Nosed Mole.

SUB-ORDER CARNIVORA.

FAMILY FELIDAE.

10. Lynx rufus, *Raf.*—Wild Cat.
11. " Canandensis, *Raf.*—Lynx.

FAMILY CANIDAE.

Sub-Family Lupinae.

12. Canis occidentalis, var. griseo-albus, *Bd.*—Wolf.
13. " latrans, *Say.*—Prairie Wolf.

Sub Family Vulpinae.

14. Vulpes fulvus, *Rich.*—Red Fox.
15. " Virginianus, *DeKay.*—Gray Fox.

FAMILY MUSTELIDAE.

Sub-Family Martinae.

x16. Mustela Penantii, *Erxl.*—Fisher.
x17. " Americana, *Turton.*—Pine Marten.
18. Putorius cicognanii, *Bd.*—Brown Weasel.
19. " Noveboracensis, *DeKay,*—White Weasel.
20. " vison, *Rich.*—Mink.
x21. Gulo luscus, *Sabine.*—Wolverine.

Sub-Family Lutrinae.

22. Lutra Canadensis, *Sab.*—Otter.

Sub Family Melinae.

23. Mephitis mephitica, *Bd.*—Skunk.
24. Taxidea Americana, *Bd.*—Badger.

FAMILY URSIDAE.

25. Procyon lotor, *Storr.*—Raccoon.
26. Ursus Americanus, *Pallas.*—Black Bear.

ORDER MARSUPIATA.

FAMILY DIDELPHIDAE.

x27. Didelphys Virginiana, *Shaw.*—Opossum.

ORDER RODENTIA.

FAMILY SCIURIDAE.

Sub-Family Sciurinae.

28. Sciurus Ludovicianus, *Custis.*—Fox Squirrel.
29. " Carolinensis, *Gm.*—Gray and Black Squirrels.

30. Sciurus Hudsonius, *Pallas.*—Red Squirrel.
31. Pteromys volucella, *Des.*—Flying Squirrel.
32. Tamias striatus, *Baird.*—Chipmunk.
×33. Spermophilus tridecem-lineatus, *Aud. & Bach.*—Striped Prairie Squirrel.
·34. Arctomys monax, *Gm.*—Woodchuck.

Sub-Family Castorinae.

×35. Castor Canadensis, *Kuhl.*—Beaver.

FAMILY MURIDAE.

Sub-Family Dipodinae.

36. Jaculus Hudsonius, *Bd.*—Jumping Mouse.

Sub-Family Murinae.

37. Mus musculus, *Linn.*—Common Mouse.
38. Hesperomys leucopus, *Wag.*—Deer Mouse.
39. " Michiganensis, *Wagner.*—Prairie Mouse.

Sub-Family Arvicolinae.

40. Arvicola riparia, *Ord.*—Meadow Mouse.
41. Fiber zibethicus, *Cuv.*—Muskrat.

FAMILY HYSTRICIDAE.

42. Erethizon dorsatus, *F. Cuv.*—Porcupine.

×9. C. cristata. The star nosed mole appears to be a very rare species within the limits of this State. I have seen but a single specimen.

×16 and ×17. N. Pennantii, and M. Americana. The Fisher and Pine Martin undoubtedly have a place in our fauna, but I have not had an opportunity of examining specimens other than hunter's skins as found in market.

×21. G. luscus. The Wolverine is seldom found in the Lower Peninsula, having been nearly exterminated.

×27. D. Virginiana. A single specimen of the Opossum was killed in Genesee county last season. The species is, however, frequently seen in the southern part of the State.

×33. S. tridecem-lineatus. The striped Prairie Squirrel is very common in the southern counties, but has not been known in the central parts of the State until within a few years past. It is gradually extending its range northward, where the timber has been removed and the land brought under cultivation.

×35. C. Canadensis. At no very remote period the Beaver was found throughout the State as is shown by the numerous remains of their dams in localities that are now deserted by them. At present their range is confined to the northern part of the Lower Peninsula, where they are found in abundance on the head waters of nearly every stream running into Lake Huron. At Alpena several hundred skins are annually brought in from Thunder Bay river and its tributaries.

ORDER RUMINANTIA.

FAMILY CERVIDAE.

x45. Alce Americanus, *Jardine.*—Moose.

x46. Rangifer caribou, *Aud. & Bach.*—Caribou.

x47. Cervus Canadensis, *Erxl.*—Elk.

48. " Virginianus, *Bodd.*—Deer.

FAMILY LEPORIDAE.

43. Lepus Americanus, *Erxl.*—Northern Hare.

44. " sylvaticus, *Bach.*—Gray Rabbit.

CLASS AVES.

ORDER RAPTORES.

FAMILY VULTURIDAE.

1. Cathartes aura, *Illiger*—Turkey Buzzard.

FAMILY FALCONIDAE.

Sub-Family Falconinae.

2. Falco anatum, *Bon.*—Duck Hawk.

3. " columbarius, *Linn.*—Pigeon Hawk.

4. " sparverius, *Linn.*—Sparrow Hawk.

Sub-Family Accipitrinae.

5. Accipiter Cooperii, *Bon.*—Cooper's Hawk.

6. " fuscus, *Gmel.*—Sharp-shinned Hawk.

Sub-Family Buteoninae.

7. Buteo borealis, *Gmel.*—Red-tailed Hawk.

8. " lineatus, *Gmel*—Red-shouldered Hawk.

---

x45. A. Americanus. The Moose is seldom seen within the limits of the State. Hunters inform me that it is still occasionally taken, but it is rapidly disappearing from its former haunts.

x46. R. Caribou. The Caribou extends its southern range to the Upper Peninsula, where it is occasionally taken by hunters.

x47. C. Canadensis. The Elk is found in abundance in the counties of Huron and Sanilac about the head waters of the Cass River. The unrelenting pursuit of hunters by means of the rifle and trap pens will soon exterminate it, unless means are taken to prevent an indiscrimate slaughter at all seasons of the year.

9. Buteo Pennsylvanicus, *Wilson.*
*10. " Swainsoni, *Bonap.*—Swainson's Buzzard.
11. Archibuteo lagopus, *Geml.*—Rough-legged Hawk.
12. " Sancti-Johannis, *Gmel.*—Black Hawk.

Sub-Family Milvinae.

13. Circus Hudsonius, *Linn.*—Marsh Hawk.

Sub Family Aquilinae.

14. Haliaetus Washingtonii, *Aud.*—Washington Eagle.
15. " leucocephalus, *Linn.*—Bald Eagle.
16. Pandion Carolinensis, *Gmel.*—Fish Hawk.

FAMILY STRIGIDAE.

Sub-Family Buboninae.

17. Bubo Virginianus, *Gmel.*—Great Horned Owl.
18. Scops Asio, *Linn.*—Mottled Owl.
19. Otus Wilsonianus, *Lesson.*—Long-eared Owl.
20. Brachyotus Cassinii, *Brewer.*—Short-eard Owl.

Sub-Family Syrninae.

21. Syrnium nebulosum, *Foster.*—Barred Owl.
22. Nyctale Acadica,—Screech Owl.

Sub-Family Nycteininae.

23. Nyctea nivea, *Daudin*—Snowy Owl.
24. Surnia ulula, *Linn.*—Hawk Owl.

ORDER SCANSORES.

FAMILY CUCULIDAE.

25. Coccygus Americanus, *Bonap.*—Yellow-billed Cuckoo.
26. " erythrophthalmus, *Bonap.*—Black-billed "

FAMILY PICIDAE.

Sub-Family Picinae.

27. Picus villosus, *Linn.*—Hairy Wood-pecker.
28. " pubescens, *Linn.*—Downy Wood-pecker.
29. Sphyrapicus varius, *Bd*—Yellow-bellied Wood-pecker.
30. Hylatomus pileatus, *Bd.*—Log Cock.

31. Centurus Carolinus, *Bon.*—Red-bellied Wood-pecker.
32. Melanerpes erythrocephalus, *Sw.*—Red-headed Wood-pecker.
33. Colaptes auratus, *Sw.*—Flicker.

ORDER INSESSORES.

SUB-ORDER STRISORES.

FAMILY TROCHILIDAE.

34. Trochilus colubris, *Linn.*—Humming Bird.

FAMILY CYPSELIDAE.

35. Chaetura pelasgia, *Steph.*—Chimney Swallow.

FAMILY CAPRIMULGIDAE.

Sub-Family Caprimulginae.

36. Antrostomus vociferus, *Bon.*—Whippoorwill.
36. Chordeiles popetue, *Bd.*—Night Hawk.

SUB ORDER CLAMATORES.

FAMILY ALCEDINIDAE.

38. Ceryle alcyon, *Boie.*—Kingfisher.

FAMILY COLOPTERIDAE.

Sub-Family Tyranninae.

39. Tyrannus Carolinensis, *Bd.*—King Bird.
40. Myiarchus crinitus, *Cab.*—Great crested Flycatcher.
41. Sayornis fuscus, *Bd.*—Pewee Fly-catcher.
x41.[a] " Sayus, *Baird*—Says Fly-catcher.
42. Contopus borealis, *Bd.*—Olive-sided Fly-catcher.
43. " virens, *Cab.*—Wood Pewee.
44. Empidonax Traillii, *Bd.*—Traill's Fly-catcher.
45. " acadicus, *Bd.*—Little Pewee.
x45.[a] " flaviventris, *Bd.*—Yellow-bellied Fly-catcher.

SUB-ORDER OSCINES.

FAMILY TURDIDAE.

Sub-Family Turdinae.

46 Turdus mustelinus, *Gmel.*—Wood Thrush.

*46.[a] Turdus Pallasii, *Cab.*—Hermit Thrush.
47. " migratorius, *Linn.*—Robin.
48. Sialia sialis, *Bd.*—Blue Bird.

Sub-Family Regulinae.

49. Regulus calendula, *Licht.*—Ruby-crowned Wren.
50. " satrapa, *Licht.*—Golden-crested "

FAMILY SYLVICOLIDAE.

Sub-Family Motacillinae.

51. Anthus Ludovicianus, *Licht.*—Tit Lark.

Sub-Family Sylvicolinae.

52. Mniotilta varia, *Vieill.*—Black and White Creeper.
53. Parula Americana, *Bon.*—Blue Yellow-backed Warbler.
54. Geothlypis trichas, *Cab.*—Maryland Yellow-throat.
55. Helminthophaga chrysoptera, *Cab.*—Golden-winged Warbler.
56. Helminthophaga ruficapilla, *Bd.*—Nashville Warbler.
57. Seiurus aurocapillus, *Sw.*—Golden-crowned Thrush.
58. " Noveboracensis, *Nuttall.*—Water Thrush.
59. " Ludovicianus, *Bon.*
60. Dendroica virens, *Bd.*—Black-throated Green Warbler.
61. " Canadensis, *Bd.*—Black-throated Blue "
62. " coronata, *Gray.*—Yellow-rumped "
63. " Blackburniae, *Bd.*—Blackburnian "
64. " castanea, *Bd.*—Bay-breasted "
65. " pinus, *Bd.*—Pine-creeping, "
66. " Pennsylvanica, *Bd.*—Chestnut-sided "
67. " aestiva, *Bd.*—Summer Yellow Bird.
68. " maculosa, *Bd.*—Black and Yellow Warbler.
69. " tigrina, *Bd.*—Cape May Warbler.
70. " discolor, *Bd.*—Prairie "
71. Myiodioctes mitratus, *Aud.*—Hooded Warbler.
72. " pusillus, *Bonap.*—Green Black-cap Fly-catcher.
73. " Canadensis, *Aud.*—Canada Fly-catcher.
74. Setophaga ruticilla, *Sw.*—Red Start.

Sub-Family Tanagrinae.

75. Pyranga rubra, *Vieill.*—Scarlet Tanager.

FAMILY HIRUNDINIDAE.

Sub-Family Hirundininae.

76. Hirundo horreorum, *Barton.*—Barn Swallow.
77. " lunifrons, *Say.*—Cliff Swallow.
78. " bicolor, *Vieill.*—White-bellied Swallow.
79. Cotyle riparia, *Boie.*—Bank Swallow.
80. " serripennis, *Bonap.*—Rough-winged Swallow.
81. Progne purpurea, *Boie.*—Purple Martin.

FAMILY BOMBYCILLIDAE.

82. Ampelis garrulus, *Linn.*—Bohemian Wax-wing.
83. " cedrorum, *Bd.*—Cedar Bird.

FAMILY LANIIDAE.

Sub-Family Laniinae.

84. Collyrio borealis, *Bd.*—Butcher Bird.
85. " excubitoroides, *Bd.*—White-rumped Shrike.

Sub-Family Vireoninae.

86. Vireo olivaceus, *Vieill.*—Red-eyed Vireo.
87. " Noveboracensis, *Bonap.*—White-eyed Vireo.
88. " flavifrons, *Vieill.*—Yellow-throated Vireo.

FAMILY LIOTRICHIDAE.

Sub-Family Miminae.

89. Mimus Carolinensis, *Gray.*—Cat Bird.
90. Harporhynchus rufus, *Cab.*—Brown Thrush.

Sub-Family Troglodytinae.

90.[a] Thriothorus Ludovicianus, *Bonap.*—Great Carolina Wren.
91. Cistothorus palustris, *Cab.*—Long-billed Marsh Wren.
92. Troglodytes aedon, *Vieill.*—House Wren.
93. " hyemalis, *Vieill.*—Winter Wren.
93.[a] " Americanus, *Aud.*—Wood Wren.

FAMILY CERTHIADAE.

94. Certhia Americana, *Bonap.*—American Creeper.

95. Sitta Carolinensis, *Gmel.*—White-bellied Nuthatch.

96. " Canadensis, *Linn.*—Red-bellied Nuthatch.

FAMILY PARIDAE.

Sub-Family Polioptilinae.

97. Polioptila caerulea, *Scl.*—Blue-Gray Fly-catcher.

Sub-Family Parinae.

98. Parus atricapillus, *Linn.*—Black cap Titmouse.

×98ª. " Carolinensis, *Aud.*—Carolina "

FAMILY ALAUDIDAE.

99 Eremophila cornuta, *Boie.*—Shore Lark.

FAMILY FRINGILLIDAE.

Sub-Family Coccothraustinae.

×69ª. Carpodacus purpureus, *Gray.*—Purple Finch.

100 Chrysomitris tristis, *Bon.*—Yellow Bird.

×100ª " pinus, *Bon.*—Pine Finch

101 Curvirostra Americana, *Wilson.*—Red Cross bill.

102 " leucoptera, *Wilson.*—White-winged Crossbill.

103 Aegiothus linaria, *Cab.*—Lesser Red Poll.

104 " canescens, *Cab.*—Mealy Red Poll.

105 Plectrophanes nivalis, *Meyer.*—Snow Bunting.

Sub-Family Spizellinae.

106. Passerculus Savanna, *Bonap*—Savannah Sparrow.

107. Poocætes gramineus, *Bd.*—Grass Finch.

108. Chondestes grammaca, *Bonap.*—Lark Finch.

109. Zonotrichia leucophrys, *Sw*—White-crowned Sparrow.

110. " albicollis, *Bonap.*—White-throated "

×110.ª Junco Oregonus, *Scl.*—Oregon Snow Bird.

111. " hyemalis, *Sclater.*—Snow Bird.

112. Spizella monticola, *Bd.*—Tree Sparrow.

113. " pusilla, *Bonap*—Field "

114. Spizella socialis, *Bonap.*—Chipping Sparrow.
115. Melospiza melodia, *Bd.*—Song "
116. " palustris, *Bd.*—Swamp "

Sub-Family Spizinae.

117. Euspiza Americana, *Bonap.*—Black-throated Bunting.
118. Guiraca Ludoviciana, *Sw.*—Rose-breasted Grosbeak.
*118.[a] " melanocephala, *Sw.*—Black-headed "
119. Cyanospiza cyanea, *Bd.*—Indigo Bird.
120. Pipilo erythrophthalmus, *Vieill.*—Chewink.

FAMILY ICTERIDAE.

Sub-Family Agelainae.

121. Dolichonyx oryzivorus, *Sw.*—Boblink.
122. Molothrus pecoris, *Sw.*—Cow Blackbird.
123. Agelaius phœniceus, *Vieill.*—Red-winged Blackbird.
124. Sturnella magna, *Sw.*—Meadow Lark.

Sub-Family Icterinae.

125. Icterus spurius, *Bonap.*—Orchard Oriole.
126. " Baltimore, *Daud.*—Baltimore Oriole.

Sub-Family Quiscalinae.

127. Scolecophagus ferrugineus, *Sw.*—Rusty Grakle.
128. Quiscalus versicolor, *Vieill.*—Crow Blackbird.

EAMILY CORVIDAE.

Sub Family Corvinae.

129. Corvus carnivorus, *Bart.*—Raven.
130. " Americanus, *Aud.*—Crow.

Sub-Family Garrulinae.

131. Cyanura cristatus, *Sw.*—Blue Jay.
132. Perisoreus Canadensis, *Bonap.*—Canada Jay.

ORDER RASORES.

SUB-ORDER COLUMBAE.

FAMILY COLUMBIDAE.

Sub-Family Columbinae.

133. Ectopistes migratoria, *Sw.*—Wild Pigeon.

Sub-Family Zenaidinae.

134. Zenaidura Carolinensis, *Bonap.*—Mourning Dove.

SUB-ORDER GALLINAE.

FAMILY PHASIANIDAE.

Sub-Family Meleagrinae.

135. Meleagris gallopavo, *Linn.*—Wild Turkey.

FAMILY TETRAONIDAE.

136. Tetrao Canadensis, *Linn.*—Canada Grouse.
137. Cupidonia cupido, *Bd.*—Prairie Chicken.
138. Bonasa umbellus, *Steph.*—Ruffed Grouse, Partridge.

FAMILY PERDICIDAE.

139. Ortyx Virginianus, *Bonap.*—Quail.

ORDER GRALLATORES.

SUB-ORDER HERODIONES.

FAMILY GRUIDAE.

139.* Grus Canadensis, *Temm.*—Sand-hill Crane.

FAMILY ARDEIDAE.

140. Ardea Herodias, *Linn.*—Blue Heron.
141. Ardetta exilis, *Gray.*—Least Bittern.
142. Botaurus lentiginosus, *Steph.*—Bittern.
143. Butorides virescens, *Bonap.*—Green Heron.
144. Nyctiardea gardeni, *Bd.*—Night Heron.

SUB-ORDER GRALLAE.

FAMILY CHARADRIDAE.

145. Charadrius Virginicus, *Borck.*—Golden Plover.
146. Aegialitis vociferus, *Cassin.*—Kill-deer.
147. " semipalmatus, *Bon.*—King Plover.
148. Squatarola Helvetica, *Cuv.*—Black-bellied Plover.

FAMILY HAEMATOPODIDAE.

149. Strepsilas interpres, *Ill.*—Turnstone.

FAMILY SCOLOPACIDAE.

150. Philohela minor, *Gray.*—Woodcock.

151. Gallinago Wilsonii, *Bonap.*—Wilson's Snipe.
152. Macrorhamphus griseus, *Leach.*—Red-breasted Snipe.
153. Tringa canutus, *Linn.*—Robin Snipe.
154. " maculata, *Vieill.*—Jack Snipe.
155. " Wilsonii, *Nuttall.*—Least Sandpiper.
×155.a " Bonapartii, *Sch.*
156. Ereunetes petrificatus, *Ill*—Semipalmated Sandpiper.
×156.a Micropalama himantopus, *Bd.*—Stilt "

Sub-Family Totaninae.

157. Gambetta melanoleuca, *Bon.*—Tell Tale.
158. " flavipes, *Bon.*—Yellow Legs.
159. Rhyacophilus solitarius, *Bonap.*—Solitary Sandpiper.
160. Tringoides macularius, *Gray.*—Spotted "
161. Actiturus Bartramius, *Bonap.*—Field Plover.
162. Limosa fedoa, *Ord.*—Marbled Godwit.

Sub-Family Rallinae.

163. Rallus Virginianus, *Linn.*—Virginia Rail.
164. Porzana Carolina, *Vieill*—Sora Rail.
165. " Noveboracensis, *Bd.*—Yellow Rail.
166. Fulica Americana, *Gm.*—Coot.
×167. Gallinula galeata, *Bonap.*—Florida Gallinule.

ORDER NATATORES.

SUB-ORDER ANSERES.

FAMILY ANATIDAE.

Sub-Family Cygninae.

168. Cygnus Americanus, *Sharpless.*—Swan.

Sub Family Anserinae.

169. Anser hyperboreus, *Pallas*—Snow Goose.
170. Bernicla Canadensis, *Boie*—Canada "

Sub-Family Anatinae.

171. Anas boschas, *Linn.*—Mallard.
172. " obscura, *Gm.*—Dusky Duck.
173. Dafila acuta, *Jenyns.*—Pintail Duck.

174. Nettion Carolinensis, *Bd*—Green-winged Teal.
175. Querquedula discors, *Steph.*—Blue-winged "
176. Spatula clypeata, *Boie.*—Spoonbill.
177. Chaulelasmus streperus, *Gray.*—Gadwall; Gray Duck.
178. Mareca Americana, *Steph.*—Baldpate; Widgeon.
179. Aix sponsa, *Boie.*—Wood Duck.

Sub-Family Fuligulinae.

180. Fulix marila, *Bd.*—Scaup Duck
181. " collaris, *Bd.*—Ring-necked Duck.
182. Aythya Americana, *Bon.*—Red head.
183. " vallisneria, *Bon.*—Canvas-back.
184. Bucephala Americana, *Bd.*—Golden Eye; Whistle Wing.
185. " albeola, *Bd.*—Butter Ball.
186. Harelda glacialis, *Leach.*—Old Wife.
187. Oidemia bimaculata, *Bd.*—Huron Scoter.

Sub-Family Erismaturinae.

188 Erismatura rubida, *Bonap.*—Ruddy Duck.

Sub-Family Merginae.

189. Mergus Americanus, *Cassin.*—Sheldrake.
190. " serrator, *Linn.*—Red-breasted Merganser.
191. Lophodytes cucullatus, *Reich.*—Hooded "

SUB-ORDER GAVIAE.

FAMILY LARIDAE.

Sub-Family Larinae.

192. Larus glaucus, *Brunn.*—Glaucous Gull.
193. " argentatus, *Brunn.*—Herring "

---

x10. B. Swainsoni. I am indebted to my friend Dr. Daniel Clark, of Flint, for an opportunity of examining a specimen of this rare buzzard, which was shot in Genesee county last summer, and is now preserved in the museum of the Flint Scientific Institute.

x41.a Sayornis Sayus, *Bd.* On the authority of Rev. Charles Fox, who shot a specimen at Owosso, Shiawassee county, July, 1863 The species in the catalogue marked 'a' were obtained at Gross Isle, Wayne Co., by Prof. Fox, and are given on his authority.

x167. G. galeata. This gallinule is frequently seen in the southern parts of the State. I have seen several specimens as far north as Saginaw Bay, and am informed by Mr. John Sharp, at the Saginaw Light-house, that it breeds in the marshes at the mouth of Saginaw River.

194. Larus Delawarensis, *Ord.*—Ring-billed Gull.
195. Chroicocephalus atricilla, *Linn.*—Laughing Gull.
196. " Philadelphia, *Lawrence.*—Bonaparte's Gull.

Sub-Family Sterninae.

197. Sterna Wilsoni, *Bonap.*—Wilson's Tern.
*197a. " frenata, *Gambel*—Least "
*198. Hydrochelidon plumbea, *Lawrence*—Black Tern.

FAMILY COLYMBIDAE.

Sub-Family Colymbinae.

199. Colymbus torquatus, *Brunn.*—Loon.

Sub-Family Podicipinae.

200. Podiceps griseigena, *Gray.*—Red-necked Grebe.
201. " cristatus, *Lath.*—Crested "
202. " cornutus, *Lath.*—Horned "
203. Podilymbus podiceps, *Lawrence.*—Pied-bill "

CLASS REPTILIA.

ORDER TESTUDINATA.

SUB-ORDER AMYDAE.

FAMILY TRIONYCHIDAE.

*1. Amyda mutica, *Fitz.*
*2. Aspidonectes spinifer, *Ag*—Soft-shelled Turtle.

FAMILY CHELYDROIDAE.

3. Chelydra serpentina, *Schw.*—Snapping Turtle.

FAMILY CINOSTERNOIDAE.

*4. Ozotheca odorata, *Ag.*
5. Thyrosternum Pennsylvanicum, *Ag.*—Musk Turtle.

FAMILY EMYDOIDAE.

*6. Graptemys geographica, *Ag.*

---

*198. H. plumbea. I shot several specimens of this beautiful tern last June, on the shore of Saginaw Bay. From the number of individuals in that vicinity I supposed it to be breeding there.

7. Graptemys LeSueurii, *Ag.*
*8. Chrysemys marginata, *Ag.*
9. Emys Meleagris, *Ag.*
*10. Nanemys guttata, *Ag.*

ORDER OPHIDIA.

FAMILY CROTALIDAE.

11. Crotalophorus tergeminus, *Holb.*—Massasauga.

FAMILY COLUBRIDAE.

*12. Eutaenia saurita, *B. & G.*—Striped Snake.
13. " sirtalis, *B & G.*—Garter Snake.
14. Nerodia sipedon, *B. & G.*—Water Snake.
15. " Agassizii, *B. & G.*
16. Regina leberis, *B. & G.*—Striped Water Snake.
*17. Heterodon platyrhinos, *Latr.*—Blowing Viper.
*18. Scotophis vulpinus, *B. & G.*
19. Ophibolus eximius, *B. & G.*—Milk Snake.
20. Bascanion constrictor, *B. & G*—Black Snake.
21. " Foxii, *B. & G.*
22. Chlorosoma vernalis, *B. & G.*—Green Snake.
23. Diadophis punctatus, *B. & G.*—Ring-necked Snake.
24. Storeria Dekayi, *B. & G.*
25. " occipito maculata, *B. & G.*

---

*1. A. mutica. This species seems to be comparatively rare. I have seen but a few specimens which would indicate that its range is confined to the southern parts of the State.

*2. A. spinifer. The common soft shell turtle is found throughout the southern half of the Lower Peninsula. It is frequently met with as far north as Genesee county, and in the streams of the eastern, as well as the western slope of the State.

*4. O. odorata. The carapace of a small turtle obtained in Oakland county I have referred to this species, but as the specimen is imperfect I may be incorrect in including the species in our fauna.

*8. C. marginata. This is the most abundant species of the Testudinata in our State. It was formerly confounded with C. picta, but was separated by Prof. Agassiz in his contributions to the Natural History of the United States. I am not aware that the latter species is found in Michigan.

*10. N. guttata. Four specimens of this beautiful species have been collected within the two years. One in Genesee county, one from Saginaw Bay, and the others from Oakland county. On comparison with a specimen from Massachusetts, they appear to be identical, the only difference noticed being the darker color of the plastron in the Michigan specimens.

CLASS BATRACHIA.

ORDER ANURA.

FAMILY BUFONIDAE.

26. Bufo Americanus, *LeConte.*

FAMILY HYLADAE.

27. Acris crepitans, *Bd.*
28. Hyla versicolor, *LeConte.*
29. " Pickeringii, *Holl.*

×30. Helocætes triseriatus, *Bd.*

FAMILY RANIDAE.

31. Rana Catesbiana, *Shaw.*—Bull Frog.
32. " fontinalis, *LeConte.*—Spring Frog.
33. " pipiens, *Gmel.*—Shad Frog.
34. " palustris, *LeConte.*—Pickerel Frog.
35. " sylvatica, *LeConte.*—Wood Frog.

ORDER URODELA.

ATRETODERA.

FAMILY AMBYSTOMIDAE.

36. Ambystoma punctatum, *Bd.*
37. " luridum, *Bd.*

×38. " laterale, *Hall.*

×12. E. saurita. This well-marked species is comparatively rare. I have seen but three or four specimens that have been collected within the limits of the State.

×17. H. platyrhinos. I have not seen this species, but give it a place in our fauna on the authority of Prof A. Sager, the able Zoologist of the former Geological corps, to whom I am indebted for many acts of kindness and encouragement.

×18. S. vulpinus. The only specimens of this species collected are from the vicinity of Saginaw Bay, where it is found in abundance. Although perfectly harmless it has the unfounded reputation, in that locality, of being venomous and is therefore much dreaded.

×30. H. triseriatus. I am not acquainted with this species, but give it a place in the catalogue on the authority of Prof. Baird.

×38. A. laterale. An immature specimen from Saginaw Bay, I have referred to this species.

×41. P. erythronota. This is a common and widely distributed species, being found throughout the State as far north as Lake Superior.

×43. I have several undetermined specimens of Necturus, some of which will probably prove to be N. maculatus.

FAMILY TRITONIDAE.

39. Diemyctylus miniatus, *Raf.*
40. " viridescens, *Raf.*

FAMILY PLETHODONTIDAE.

×41. Plethodon erythronota, *Bd.*
42. " cinereus, *Tsch.*

Tremadotera.

×43. Necturus lateralis, *Bd.*

CLASS GASTEROPODA.

MOLLUSCA.

FAMILY HELICIDAE.

1. Helix albolabris, *Say.*
×2. " alternata, *Say.*
3. " arborea, *Say.*
4. " chersina, *Say.*
5. " concava, *Say.*
6. " clausa, *Say.*
7. " exoleta, *Binney.*
8. " electrina, *Gould.*
9. " elevata, *Say.*
10. " fallax, *Say.*
11. " fraterna, *Say.*
12. " fuliginosa, *Griffith.*
13. " hirsuta, *Say.*
×14. " hydrophyla, *Ing.*
15. " inflecta, *Say.*
17. " identata, *Say.*
18. " inornata, *Say.*
19. " ligera, *Say.*
20. " labyrinthica, *Say.*
21. " limatula, *Ward.*
22. " lineata, *Say.*

23. Helix minuscula, *Binney.*
24. " monodon, *Rack.*
25. " multilineata, *Say.*
26. " palliata, *Say.*
27. " perspectiva, *Say.*
28. " profunda, *Say.*
x29. " pulchella, *Miller.*
30. " Sayii, *Binney.*
31. " solitaria, *Say.*
32. " striatella, *Anth.*
33. " thyroides, *Say.*
34. " tridentata, *Say.*
35. Bulimus marginatus, *Say.*
36. Achatina lubrica, *Mull.*
37. Succinea campestris, *Say.*
38. " avara, *Say.*
39. " ovalis, *Say.*
40. " vermetus, *Say.*
41. " obliqua, *Say.*
42. Pupa pentodon, *Say.*
43. " armifera, *Say.*
44. " contracta, *Say.*
45. Vertigo Gouldii, *Binn.*
46. " ovata, *Say.*
47. " simplex, *Gld.*

FAMILY AURICULIDAE.

48. Carychium exiguum, *Say.*

FAMILY LIMNEIDAE.

49. Planorbis armifera, *Say.*
50. " bicarinatus, *Say.*
51. " campanulatus, *Say.*
x52. " deflectus, *Say.*
53. " exacutus, *Say.*
54. " lentus, *Say.*
55. " parvus, *Say.*

56. Planorbis trivolvis, *Say.*
*57. " truncatus, *Nobis.*
58. Physa heterostropha, *Say.*
59. " elongata, *Say.*
60. " Hildrethiana, *Lea.*
61. " vinosa, *Gld.*
62. Limnea appressa, *Say.*
63. " columella, *Say.*
64. " caperata, *Say.*
65. " desidiosa, *Say.*
66. " elodes, *Say.*
67. " gracilis, *Say.*
68. " jugularis, *Say.*
69. " modicellus, *Say.*
70. " reflexa, *Hald.*
71. " umbilicata, *Adams.*
72. " umbrosa, *Say.*
73. " pallida, *Adams.*
74. Ancylus fuscus, Adams.
75. " paralellus, *Hald.*
76. " tardus, *Say.*

FAMILY MELANIADAE.

77. Melania Virginica, *Say.*
78. " depygis, *Say.*
79. " Niagarensis, *Lea.*
80. " neglecta, *Anth.*
81. " livescens, *Menka.*
82. " pulchella, *Anth.*

FAMILY PALUDINIDAE.

83. Valvata sincera, *Say.*
84. " tricarinata, *Say.*
*85. " humeralis, *Say.*
86. Paludina decisa, *Say.*
87. " integra, *Say.*
88. " isogona, *Say.*

x89. Paludina obesa, *Lewes.*
90. " ponderosa, *Say.*
91. " rufa, *Hald.*
92. Amnicola grana, *Gould.*
93. " lapidaria, *Say.*
94. " pallida, *Hald.*

CLASS ACEPHALA.

SIPHONIDA.

FAMILY CYCLADIDAE.

95. Sphærium occidentale, *Prime.*
96. " partumeia, *Say.*
97. " solidulum, *Prime.*

---

x2. H. alternata. This seems to be the most widely distributed mollusk in the State, being found everywhere as far north as Lake Superior.

x14. H. hydrophyla. I am indebted for this species to Mr O. A. Currier, of Grand Rapids, who has made extensive collections in the Grand River Valley, and has a valuable cabinet of native shells to which he has given me free access, thus materially facilitating my labors in this department.

x29. H. pulchella. Mr. Albert D. White, who has rendered me valuable assistance in collecting Zoological specimens, has furnished a suite of the Helicidae from Ann Arbor, containing this species. It is found there in abundance.

x52. P. deflectus. This species is added to the catalogue on the authority of Mr. Currier.

x57. P. truncatus, nobis. Shell sub-orbicular, color light chestnut; the right side deeply umbilicated, the concavity bordered by an obtuse carina; the volutions seen from this side are scarcely more than two; left side truncated, presenting a flat surface extending across all the whorls, the suture being marked by a minute raised line, which likewise extends around the edge of the truncation; the space between the volutions of this raised line, as well as the entire body of the shell, is beautifully marked with delicate longitudinal lines, which are crossed by the minute, raised, transverse lines of growth; the longitudinal lines are scarcely distinguishable without the aid of a microscope; whorls on left side four or five; aperture ovate, widest on the right side, which extends beyond the general plane of that side of the shell; the lip on the left side is straight for a short distance from the body whorl, and in a line with the truncated plane, at the outer edge of which it forms an angle, marked on the inner surface by a slight groove, corresponding to the raised line separating the whorls on the outside; lip thin, slightly thickened by a bluish-white callus, bordered on the inner edge by a purplish band; the longitudinal lines, as well as the transverse lines of growth, are distinctly seen within the aperture. Measurements, .6—.35. Hab. Saginaw Bay. In a few specimens the growth of the whorls has not been in the same plane, leaving a slightly projecting turreted spire on the left side.

x85. V. humeralis. Grand River. Mr. Currier's cabinet.

x89. P. obesa. Grand River Valley. Cabinet of Mr. Currier.

98. Sphærium striatinum, *Lam.*
99. " sulcatum, *Lam.*
100. Pisidium abditum, *Hald.*
101. " compressum, *Prime.*
102. " ventricosum, *Prime.*
103. " Virginicum, *Bgt.*

ASIPHONIDA.

FAMILY UNIONIDAE.

104. Unio alatus, *Say.*
105. " asperrimus, *Lea.*
106. " bullatus, *Raf*
107. " coccineus, *Hld.*
108. " complanatus, *Lea.*
109. " coelatus, *Con.*
*110. " cariosus, *Say.*
111. " circulus, *Lea.*
112. " ellipsis, *Lea.*
113. " elegans, *Lea.*
114. " gibbosus, *Bar.*
115. " gracilis, *Bar.*
*116. " glans, *Lea.*
117. " Hildrethianus, *Lea.*
118. " iris, *Lea.*
119. " lapillus, *Say*
120. " lævissimus, *Lea.*
*121. " luteolus, *Lam.*
122. " ligamentinus, *Lam.*
123. " multiradiatus, *Lea.*
124. " Novi-Eboraci, *Lea.*
125. " nasutus, *Say.*
126. " occidens, *Lea.*
127. " plicatus, *Say.*
128. " perplexus, *Lea.*
129. " penitus, *Con.*
130. " pressus, *Lea.*

131. Unio phaseolus, *Hild.*
132. " rectus, *Lam.*
133. " rubiginosus, *Lea.*
×134. " leprosus, *Nobis.*
135. " subrotundus, *Lea.*
136. " Schoolcraftensis, *Lea.*
137. " spatulata, *Lea.*
138. " subovatus, *Lea.*
139. " tenuissimus, *Lea.*
140. " trigonus, *Lea.*
141. " triangularis, *Bar.*
142. " undulatus, *Bar.*
143. " verrucosus, *Bar.*
144. " ventricosus, *Bar.*
145. Alasmodon rugosa, *Bar.*
146. " marginata, *Say.*
147. " deltoides, *Lea.*
×148. Anodonta Benedictii, *Lea.*
149. " cataracta, *Say.*
150. " edentula, *Lea.*

×110. U. cariosus. I give this species on the authority of Prof. Sager.

×116. U. glans. This shell was found in the Clinton River, at Pontiac, Oakland county, by Mr. John A. McNiel, an enthusifstic and indefatigable collector of shells, residing at Grand Rapids.

×121. U. luteolus. This bivalve presents a great variety in form and appearance, and is found in every part of the State. Among the collections are several well marked varieties that may prove to be distinct species on further examination.

×134. U. leprosus, nobis. Shell, thick, oblong, transverse, very inequilateral, compressed towards the basal margin; posterior extremity rounded, nearer the basal than the dorsal margin; anterior extremity sub-truncate; beaks slightly elevated; anterior lunule distinct, extending between the beaks; umbonal slope rounded, prominent; basal and hinge margins nearly parallel; epidermis reddish brown, somewhat roughened by the lines of growth; cardinal teeth massive, prominent; lateral teeth long, elevated, slightly curved; nacre white iridescent, with dark blotches towards the beaks, roughened by numerous pearlaceous tubercles; anterior cicatrices large, deep; posterior cicatrices large, confluent, slightly impressed; dorsal cicatrices deeply impressed, situated in the shallow cavity of the beaks. Diam. 1.56. Length, 2.65. Breadth, 6, Hab. Huron River Livingston county.

×148. There are undoubtedly several additional species of the genus Anodonta, in the collections already made, which have not been determined, some of which may prove to be undescribed. I am indebted to Mr. Currier's cabinet for several species in the catalogue of this genus.

151. Anodonta fluviatilis, *Lea.*
152. " Ferrussaciana, *Lea.*
153. " Footiana, *Lea.*
154. " imbecilis, *Say.*
155. " modesta, *Lea.*
156. " ovata, *Lea.*
157. " plana, *Lea.*
158. " pallida, *Anth.*
159. " Pepiniana, *Lea.*
160. " Shafferiana, *Lea.*
161. " subcylindracea, *Lea.*

PART III.

BOTANY.

CHAPTER IX.

CATALOGUE OF PHÆNOGAMOUS AND ACROGENOUS PLANTS FOUND GROWING WILD IN THE LOWER PENINSULA OF MICHIGAN AND THE ISLANDS AT THE HEAD OF LAKE HURON.

During the season of 1859, no special botanical assistant was connected with the survey. As the work of 1860, was to extend into portions of the State less known to the botanist, Mr. N. H. Winchell was selected to accompany the exploring party in the special capacity of botanical collector and assistant, and the following catalogue has been drawn up by his hands.

The following are the sources from which the materials for this catalogue have been derived:

1. The observations of the geological parties in 1859 and 1860.

2. The catalogue published by Dr. Wright in the Geological Report of 1838.

3. The University Herbarium which contains many plants collected after the publication of Dr. Wright's Calalogue. A list of these plants was made out at my request, and the whole collection arranged by Mr. E. E. Baldwin.

4. The catalogue prepared by W. D. Whitney, of plants observed in the Lake Superior Land District, and published in Foster and Whitney's Report, vol. ii.

5. The notes of Miss Mary Clark, of Ann Arbor, an enthusiastic botanist and collector from various parts of the State.

6. A collection of plants made in the neighborhood of Fort Gratiot, near the foot of Lake Huron, by Mr. E. P. Austin, Assistant on the Coast Survey of the lakes.

7. Observations made by the writer during several years past in the vicinity of Ann Arbor.

8. A very few species have been admitted on the authority of Gray's Manual of Botany.

The catalogue shows, except in the case of very common plants, every locality where each species was noted, and, affixed to this, the date, provided the plant was seen in flower. Such plants as are common to this list and Dr. Wright's, have their localities designated, in a general way, by initials corresponding to the four quarters of the Lower Peninsula, thus: "S. E., (Wright)," "S. W., (Wright)," &c. All other localities are definitely stated, and the authority, if other than our own observations, follows in parenthesis. The corrections of nomenclature within the space of 20 years have converted many of Dr. Wright's names into synonyms, which are made to follow the modern name thus: Hepatica triloba, Chaix, *(H. Americana—W.)*

For the purpose of convenient reference, as well as economizing space, the common names of most of the species have been placed in the left hand margin opposite the scientific names.

LIST OF PLANTS.

RANUNCULACEÆ.

Virgin's Bower. Clematis virginiana, L. (*C. virginica—W.*)
Emmet Co.; Ann Arbor, (Wright.)

Many Cleft Anemone. Anemone multifida, DC.
Mouth Saginaw River, 14 June; Mackinac.

Long Fruited Anemone. Anemone cylindrica, Gray.
Ann Arbor; Pigeon R., 18 June.

Tall Anemone. Anemone virginiana, L.
Drummond's I.; Ann Arbor, (Wright); Ft. Gratiot, (Austin).

Pennsylvanian Anemone. Anemone Pennsylvanica, L. (*A. acontifolia.—W.*)
Shore Saginaw B.; Ann Arbor; Ft. Gratiot.

Wind Flower. Anemone nemorosa, L.
Ann Arbor, very common; Ft. Gratiot.

Round Lobed Hepatica. Hepatica triloba, Chaix. (*H. americana.—W*)
Ann Arbor; very common.

Sharp Lobed Hepatica. Hepatica acutiloba, DC.
Ann Arbor, very common; S. W. (Wright.)

Rue Anemone. Thalictrum anemonoides, Michx.
Ann Arbor.

Early Meadow Rue. Thalictrum dioicum, L.
Ann Arbor.

Meadow Rue. Thalictrum Cornuti, L.
Ann Arbor; Stone I., Saginaw B.; Sulphur I., north of Drummond's; Ft. Gratiot.

White Water-Crowfoot. Ranunculus aquatilis, L.
var. divaricatus.
Ann Arbor; Middle I., Lake Huron, 9 July; Ft. Gratiot.

Yellow Water-Crowfoot. Ranunculus Purshii, Richards. (var. *fluviatilis*—Univ. Herb.)
Ann Arbor; Ft. Gratiot.

Spearwort. Ranunculus Flammula, L.

Creeping Spearwort. var. reptans
St. Mary's R., 31 July; S. E. (Univ. Herb.); L. of Lilies, (Miss Clark.)

Ranunculus rhomboideus, Goldie.
"Prairies, Michigan," (Gray.)

Small Flowered Crowfoot. Ranunclus abortivus, L.
Ann Arbor, common; Stone I., Saginaw Bay; Ft. Gratiot.
var. micranthus.
Ann Arbor; Drummond's I.

Cursed Crowfoot. Ranunculus sceleratus, L.
Ann Arbor; St. Helena I., L. Mich., 10 Aug.; Ft. Gratiot.

Hooked Crowfoot. Ranunculus recurvatus, Poir.
Ann Arbor; Ft. Gratiot.

Bristly Crowfoot. Ranunculus Pennsylvanicus, L.
S. W. (Wright); Ann Arbor, (Miss Clark.)

Early Crowfoot. Ranunculus fascicularis, Muhl.
Ann Arbor, common.

Creeping Crowfoot. Ranunculus repens, L.
Ann Arbor; Pigeon R., 18 June.

Buttercups. Ranunculus acris, L.
Mackinac, 19 July; Saut St. Marie, abundant as well as at Mackinac; Ft. Gratiot.

Marsh Marigold. Catha palustris, L.
Ann Arbor, Sturgeon Pt., L. Huron, very large, deeply crenate leaves.

Spreading Globe-flower. Trollius laxus, Salisb.
"Deep swamps, Mich." (Gray.)

Three leaved Goldthread. Coptis trifolia, Salisb.
S. E. (Wright); Mont Lake, (Miss Clark.)

Wild Columbine. Aquilegia Canadensis, L.
Ann Arbor; shore of Saginaw B.; Drummond's I.; Ft. Gratiot.

Tall Larkspur. Delphinium exaltatum, Ait.
"Rich soil," (Gray.)

Orangeroot. Hydrastis Canadensis, L.
Ann Arbor, (Wright.)

Red Baneberry. Actaea spicata, L. var. rubra, Michx. (*A. rubra.—W.*)
Shore of Saginaw Bay; Drummond's I.; Ann Arbor; (Miss Clark).

White Baneberry Cohosh. var. alba, Michx. (*A. alba.—W.*)
Ann Arbor; Pt. au Chene, L. Mich.

Black Snakeroot. Cimcifuga racemosa, Ell.
S. E. (Univ. Herb).

MAGNOLIACEÆ.

Tulip-tree, Liriodendron Tulipifera, L.
Ann Arbor.

ANONACEÆ.

Common Papaw. Asimina triloba, Dunal.
Monroe Co.; Farmington; Ann Arbor, (Miss Clark).

MENISPERMACEÆ.

Canadian Moonseed. Menispermum Canadense, L
S. W. (Wright); Ann Arbor, (Miss Clark).

BERBERIDACEÆ.

Blue Cohosh, Pappoose-root. Caulophyllum thalictroides, Michx.
Ann Arbor, (Miss Clark).

Mandrake, May-Apple. Podophyllum peltatum, L.
Ann Arbor, very common; shore of Saginaw Bay; Ft. Gratiot.

Twin-leaf. Jeffersonia diphylla, Pers.
Ann Arbor, (Miss Clark).

CABOMBACEÆ.

Water-shield. Brasenia peltata, Pursh.
S. E. (Univ. Herb).

NYMPHÆACEÆ.

Sweet-scented Water-Lily. Nymphæa odorata, Ait.
Ann Arbor; Ft. Gratiot.

Yellow Pond Lily Spatter-dock. Nuphar advena, Ait.
Saginaw B., common, 15 June; St. Mary's R., in flower July 31; Ann Arbor, (Miss Clark); F. Gratiot.

Yellow Pond Lily Spatter-dock. Nuphar Kalmiana, Pursh.
Saginaw B., 15 June; S. W. (Wright).

SARRACENIACEÆ.

Pitcher-plant. Sarracenia purpurea, L.
Ann Arbor; near "sitting rabbit," 17 Aug.

PAPAVERACEÆ.

Blood-root. Sanguinaria Canadensis, L.
Ann Arbor; St. Joseph's I.

FURMARIACEAE.

Climbing Fumigory. Adlumia cirrhosa, Raf.
Middle I., L. Huron, 9 July; Grand Rapids, (Miss Clark).

Dutchman's Breeches. Dicentra Cucullaria, DC.
Detroit, (Austin).

Squirrel Corn. Dicentra Canadensis, DC.
Cape Ipperwash, C. W., (Austin). Will undoubtedly be found within our limits.

Golden Corydalis. Corydalis aurea, Willd.
Middle I., L. Huron, 9 July; Drummond's I, 23 July.

Pale Corydalis. Corydalis glanca, Pursh.
Sanilac, (Austin); Drummond's I., 23 July, has the spur and lower part of corolla pale red, and the upper part, with the tips of the petals, yellow; less common than the preceding, both preferring the vicinity of new clearings.

CRUCIFERÆ.

Water cress. Nasturtium officinale, R. Br.
Northfield, Ann Arbor, (Miss Clark).

Marsh cress. Nasturtium palustre, DC.
Ann Arbor; shore of Saginaw Bay.

Nasturtium amphibium, R. Br.
S. Michigan, (Wright).

Lake cress. Nasturtium lacustre, Gray. (*N. natans*—*W.*)
S. E. (Univ. Herb).

Horseradish. Nasturtium Armoracia, Fries.
Ann Arbor; Pigeon river, 18 June.

Toothwort, Pepper root. Dentaria diphylla, L.
Ann Arbor.

Toothwort, Pepper-root. Dentaria laciniata, Muhl.
Ann Arbor; N. E. (Univ. Herb).

Spring cress. Cardamine rhomboidea, DC.
Ann Arbor.
var. purpurea, Torr.
Ann Arbor.

Cuckoo-flower. Cardamine pratensis, L.
Ann Arbor; S. W. (Wright); Livingston Co., (Miss Clark).

Common Bitter cress. Cardamine hirsuta, L.
St. Helena I., L. Mich., 20 Aug.; S. W. (Wright); Ann Arbor, (Miss Clark).
var. Virginica, Michx.
Ann Arbor, (Miss Clark).

Rock cress. Arabis lyrata, L.
Sand Pt. Saginaw B., 17 June; S. E. (Wright); Mont Lake, (Miss Clark). The specimens seen at Sand Pt. were the variety (*Sisymbrium arabidoides*, *Hook.*) peculiar to "Upper Michigan and northward."

Rock cress. Arabis hirsuta, Scop. (*A. sagittata.*—*W.*)
Middle I., L. Huron, 8 July; S. E. (Wright).

Rock cress. Arabis lævigata, DC.
Alpena; S. Michigan, (Wright).

Sickle pod. Arabis Canadensis, L.
S. E. (Wright).

Tower mustard. Turritis glabra, L.
Gros cap, L. Mich., 18 Aug.

Turritis stricta, Graham.
Stone I., Saginaw B., 16 June.

Turritis brachycarpa, Torr. & Gray.
Ann Arbor; Alpena; Ft. Gratiot, (Gray).

Winter cress, Yellow rocket. Barbarea vulgaris, R. Br.
Thunder B. Is.; St. Helena I., L. Mich., in blossom here 20 Aug., as it was at Thunder B. July 7th.

Hedge Mustard. **Sisymbrium officinale, Scop.**
Ann Arbor.

Tansy Mustard. **Sisymbrium canescens, Nutt.**
Shore of L. Mich.

White Mustard. **Sinapis alba, L.**
Ann Arbor.

Field Mustard, Charlock. **Sinapis arvensis, L.**
Ann Arbor.

Black Mustard. **Sinapis nigra, L.**
Ann Arbor.

Whitlow-grass. **Draba arabisans, Michx.**
"Upper Michigan," (Gray).

Whitlow-grass. **Draba nemorosa, L.**
Ft. Gratiot, (Gray).

Wild pepper-grass. **Lepidium Virginicum, L.**
Ann Arbor; Saginaw Bay, 14 June.

Lepidium intermedium, Gray.
N. W. (Gray).

Shepherd's purse **Capsella Bursa-pastoris, Moench.**
Ann Arbor; Saut St. Marie, 30 July. Abundant everwhere.

American sea-rocket. **Cakile Americana, Nutt.**
Pt. au Chene, L. Mich., 18 Aug.; frequently seen on sandy beaches; rarely seen with both joints of the pod containing a perfect seed.

CAPPARIDACEÆ.

Polanisia. **Polanisia graveolens, Raf.**
S. Michigan, (Wright).

VIOLACEÆ.

Round-leaved Violet. **Viola rotundifolia, Michx.**
Sugar Island.

Sweet White Violet. **Viola blanda, Willd.**
Ann Arbor; North shore L. Mich.

Common Blue Violet. **Viola cucullata, Ait.**
Ann Arbor; Saginaw B.; Drummond's I.; Ft. Gratiot.

Hand-leaf Violet. **var. palmata.**
Ann Arbor, (Miss Clark).

Arrow-leaved Violet. **Viola sagittata, Ait.** (*V. ovata—W.*)
Ann Arbor; Detroit, (Miss Clark).

Bird-foot Violet. **Viola pedata, L.**
Ann Arbor.

Long-spurred Violet. **Viola rostrata, Pursh.**
Ann Arbor, common in May.

American Dog Violet. **Viola Muhlenbergii, Torr.**
Ann Arbor.

Pale violet. **Viola striata, Ait.**
Ann Arbor.

Canada Violet. **Viola Canadensis, L.**
Ann Arbor; Emmet Co., 26 Aug.

Downy Yellow Violet. **Viola puvescens, Ait.**
Ann Arbor; Ft. Gratiot; Emmet Co.; common.
var. eriocarpa, Nutt.
Ann Arbor; Emmet Co.; common.

CISTACEÆ.

Frostweed. **Helianthemum Canadense, Michx.**
Ann Arbor; Mouth Saginaw R., 14 June.

Hudsonia. **Hudsonia tomentosa, Nutt.**
S. Michigan, (Univ. Herb).

Pin-weed. **Lechea major, Michx.**
S. Mich., (Wright).

DROSERACEÆ.

Round-leaved Sundew. **Drosera rotundifolia, L.**
Mouth Saginaw R.; Saut St. Marie, 23 July.

Drosera longifolia, L.
S. Michigan, (Wright).

PARNASSIACEÆ.

Grass of Parnassus. **Parnassia palustris, L.**
Ann Arbor; Drummond's I., 22 July, none of the leaves heart-shaped, though the sterile filaments were about 9.

Grass of Parnassus. **Parnassia Caroliniana, Michx.** (*P. Americana*—*W.*)
North shore of L. Mich., 17 Aug.; S. Mich. (Wright).

HYPERICACEÆ.

Giant St. Johns-wort. **Hypericum pyramidatum, Ait.** (*H. Acyroides*—*W.*)
S. Mich. (Wright); Ft. Gratiot.

Hypericum Kalmianum, L.
Ft. Gratiot, Gros Cap, L. Mich., 18 Aug.; Port Huron, "marshy margin of river," (Miss Clark); S. Mich. (Wright).

Shrubby St. Johns-wort. **Hypericum prolificum, L.**
Drummond's I., 22 July; S. W. (Wright); Ann Arbor.

Hypericum corymbosum, Muhl. (*H. punctatum*—*W.*)
Ann Arbor; Ft. Gratiot, S. Mich. (Wright).

Common St. Johns-wort. **Hypericum perforatum, L.**
Ann Arbor, (Miss Clark).

Hypericum ellipticum, Hook.
Ann Arbor, (Miss Clark).

Hypericum mutilum, L. (*H. parviflorum.*—*W.*)
S. W. (Wright); "Elmwood," (Miss Clark).

Hypericum Canadense, L.
Ann Arbor; Ft. Gratiot; Sulphur I., north of Drummond's, 8 Aug.; S. W. (Wright).

Marsh St. John's-wort. Elodea Virginica, Nutt. (*Hypericum Virginicum—W.*)
S. Michigan, (Wright).

Marsh St. John's-wort. Elodea petiolata, Pursh.
Grosse Isle, (Miss Clark).

CARYOPHYLLACEÆ.

Common Soapwort, Bouncing Bet. Saponaria officinalis, L.
Ann Arbor, S. Michigan, (Wright).

Cow-Herb. Vaccaria vulgaris, Host. (*Saponaria vaccaria—W.*)
S. Michigan, (Wright).

Starry Campion. Silene stellata, Ait.
S. Michigan, (Wright).

Fire Pink, Catchfly. Silene Virginica, L.
S. Mich. (Univ. Herb).

Wild Pink. Silene Pennsylvanica, Michx.
Mont Lake, (Miss Clark).

Sleepy Catchfly. Silene antirrhina, L.
Mouth of Saginaw River, 14 June; S. E. (Wright).

Night-flowering Catchfly. Silene noctiflora, L.
Port Huron, (Miss Clark).

Corn-Cockle. Agrostemma Githago, L.
Ann Arbor.

Sandwort. Alsine Michauxii, Fenzl. (*Arenaria stricta—W.*)
S. Mich. (Wright).

Thyme-leaved Sandwort. Arenaria serpyllifolia, L.
Ann Arbor; Mackinac, 19 July, common.

Mœhringia. Mœhringia lateriflora, L. (*Arenaria lateriflora—W.*)
S. Mich. (Wright).

Common Chickweed. Stellaria media, Smith.
Ft. Gratiot, S. Mich. (Wright).

Stitchwort. Stellaria longifolia, Muhl.
Ann Arbor; Ft. Gratiot; Bruce Mine, Ca., 26 July.

Long-stalked Stitchwort. Stellaria longipes, Goldie.
Gros Cap, L. Mich., 18 Aug., abundant in pure sand.

Mouse-ear Chickweed. Cerastium vulgatum, L.
Ann Arbor; Mackinac, 19 July.

Field Chickweed. Cerastium arvense, L.
S. Michigan, (Univ. Herb.)

Larger Mouse-ear Chickweed. Cerastum viscosum, L.
Ann Arbor; Ft. Gratiot; Willow-Creek, 20 June.

Corn Spurrey. **Spergula arvensis, L.**
S. Mich. (Wright).

Forked Chick-weed. **Anychia dichotoma, Michx.**
S. W. (Wright).

Carpet-weed. **Mollugo verticillata, L.**
Ft. Gratiot; S. Mich. (Wright).

PORTULACACEÆ.

Common Purslane. **Portulaca oleracea, L.**
Ann Arbor; common.

Spring Beauty. **Claytonia Virginica, L.**
Ann Arbor; Mackinac, (Whitney).

MALVACEÆ.

Common Mallow. **Malva rotundifolia, L.**
Ann Arbor.

Velvet-Leaf. **Abutilon Avicennae, Gaertn.**
Ann Arbor.

Bladder Ketmia. **Hibiscus Trionum, L.**
Ann Arbor.

TILIACEÆ.

Basswood, Linden. **Tilia Americana, L.** (*T. glabra.—W.*)
Ann Arbor; Drummond's I.; Emmet Co.; Antrim Co.; Pt. au Chene, L. Mich. The Basswood is of frequent occurrence throughout the Southern Peninsula, nowhere forming, however, a considerable portion of the forest growth. It is most common along the inland lakes of Emmet and Antrim counties, where it attains a largê size, comparing favorably with the surrounding Elms, Beaches and Birches, in the beauty of its foliage and symmetry of its trunk.

LINACEÆ.

Wild Flax. **Linum Virginianum, L.**
S. Mich., (Wright).

Larger Yellow Flax. **Linum Boottii, Planchon.**
S. Michigan, (Univ. Herb).

Common Flax. **Linum usitatissimum, L.**
S. Mich., (Wright).

OXALIDACEÆ.

Violet Wood-Sorrel. **Oxalis violacea, L.**
S. E. (Univ. Herb).

Yellow Wood-sorrel. **Oxalis stricta, L.**
Ann Arbor.

Oxalis corniculata, L.
S. Mich. (Univ. Herb.)

GERANIACEÆ.

Wild Cranesbill. **Geranium maculatum, L.**
Ann Arbor, common; S. shore of Saginaw B., common.

Carolina Cranesbill. **Geranium Carolinianum, L.**
Drummond's I.; Alcona Co., 1 July. Occurs sparingly throughout the northern counties.

Herb Robert. **Geranium Robertianum, L.**
Stone I., Saginaw B., 16 June; S. Mich. (Wright); Middle I., L. Huron; Drummond's I; Mackinac. More common than the preceding.

BALSAMINACEÆ.

Pale Touch-me-not. **Impatiens pallida, Nutt.**
Bruce Mine, Ca., 27 July; S. E. (Wright); Sugar I., abundant, 1 Aug.

Spotted Touch-me-not. **Impatiens fulva, Nutt.**
Ann Arbor; Sugar I., 31 July; Branch L., Antrim Co. The prevailing species.

RUTACEÆ.

Northern Prickly Ash, Toothache tree. **Zanthoxylum Americanum, Mill.**
Ann Arbor; Stone I., Saginaw B.

Shrubby Trefoil, Hop-tree. **Ptelea trifoliata, L.**
S. Mich., (Wright).

ANACARDIACEÆ.

Staghorn Sumach. **Rhus typhina, L.**
Ann Arbor; Stone I., Saginaw B., 16 June; Emmet Co.; Grand Traverse Co.; S. W. (Wright).

Smooth Sumach. **Rhus glabra, L.**
Ann Arbor; Stone I., Saginaw B.; N. shore of L. Mich.; S. W. (Wright).

Dwarf Sumach. **Rhus copalina, L.**
S. W. (Wright); Detroit, (Miss Clark).

Poison Sumach or Dogwood. **Rhus venenata, DC.**
S. Mich. (Wright).

Poison Ivy. Poison Oak. **Rhus Toxicodendron, L.**
Ann Arbor; Stone I., Saginaw B., 16 June; common in the counties bordering on L. Huron; Sault St. Marie, common; less common on L. Mich.

Rhus radicans, L.
Bear Creek, Emmet Co.; S. E. (Wright).

Fragrant Sumach **Rhus aromatica, Ait.**
Dover, (Miss Clark).

VITACEÆ.

Summer Grape. **Vitis æstivalis, Michx.**
Ann Arbor; S. Mich. (Wright).

Winter or Frost Grape. **Vitis cordifolia, Michx.**
Ann Arbor; Drummond's I.; Stone I., Saginaw B.; Sand dunes of Emmet Co., its vines covering the surface of the sand in abundance.
var riparia, (*V. riparia*—*W.*)
S. E. (Wright).

Virginian Creeper. **Ampelopsis quinquefolia, Michx.**
Charity Is., Sag. B., 27 June; Ann Arbor.

RHAMNACEÆ

Buckthorn. **Rhamnus alnifolius, L'Her.** (*R. franguloideus*—*W.*)
S. E. (Wright).

New Jersey Tea. **Ceanothus Americanus, L.**
Ann Arbor; Ft. Gratiot; Sand Pt., Saginaw B., 17 June; Emmet Co.

CELASTRACEÆ.

Wax-work. Climbing Bitter-sweet. **Celastrus scandens, L.**
Ann Arbor; S. W. (Wright).

Burning-Bush. Waahoo. **Euonymus atropurpureus, Jacq.**
S. E. (Wright).

Strawberry Bush **Euonymus Americanus, L. var. obovatus, Torr. & Gray.** (*E. obovatus*—*W.*)
S. W. (Wright); Ann Arbor, (Miss Clark).

SAPINDACEÆ.

American Bladder-nut. **Staphylea trifolia, L.**
S. W. (Wright); Ann Arbor, (Miss Clark).

Fetid or Ohio Buckeye. **Aesculus glabra, Wild.**
S. Michigan, (Wright).

Striped Maple. **Acer Pennsylvanicum, L.**
Alcona Co., (most southern known limit of its range in the State); common at False Presqu' Isle, and northward, a small slender tree, the largest specimens seen measuring 5 inches in diameter, 3 feet from the surface.

Mountain Maple. **Acer spicatum, Lam.**
Alcona Co., 1 July; False Presqu' Isle, common, and northward. This is the prevailing species on the high lands of Drummond's, St. Joseph's and Sugar Islands; smaller than the last.

Sugar Maple. **Acer saccharinum, Wang.**
Ann Arbor; Mackinac, common, but the only species seen on the island! ; Emmet, Antrim and Leelanaw counties, forming here a conspicuous and important portion of the forest timber. Common throughout the State.

Black Sugar Maple. var. nigrum, (*A. nigrum.*—*W.*)
Ann Arbor.

White or Silver Maple. **Acer dasycarpum, Ehrhart.** (*A. eriocarpum.*—*W.*)
Ann Arbor.

Red or Swamp Maple. **Acer rubrum; L.**
Ann Arbor; Bruce Mine, Ca.; Branch L., Antrim Co.

Ash-leaved Maple, Box-Elder. **Negundo aceroides, Moench.**
S. Mich., (Univ. Herb).

POLYGALACEÆ.

Milkwort. **Polygala sanguinea, L.** (*P. purpurea.—W.*)
S. W. (Wright); Ann Arbor, (Miss Clark).

Milkwort. **Polygala cruciata, L.**
S. Mich., (Wright).

Polygala verticillata, L.
Ann Arbor; S. W. (Wright).

Seneca Snake-root. **Polygala Senega, L.**
Ann Arbor; shore of Saginaw B.; Drummond's I.; Sugar I.; Sault Ste Marie; North shore of L. Mich.

Polygala polygama, Walt.
Ft. Gratiot; S. Mich. (Univ. Herb.)

Flowering wintergreen. **Polygala paucifolia, Willd.**
Ann Arbor; Drummond's I.
var. alba. Eights.
S. Mich. (Wright).

LEGUMINOSÆ.

Wild Lupine. **Lupinus perennis, L.**
Ann Arbor; mouth of Saginaw R.

Red Clover. **Trifolium pratense, L.**
Ann Arbor; Pigeon river, 18 June; Presqu' Isle; Drummond's I.; Grand Traverse Co. Common everywhere.

White Clover. **Trifolium repens, L.**
Ann Arbor; Bois Blanc I., 15 July; Saut St. Marie; Emmet Co., woodlands.

Sweet Clover, White Melilot. **Melilotus alba, Lam.**
Ann Arbor; Pine L., Emmet Co., 23 Aug.

Lead Plant. **Amorpha canescens, Nutt.**
Western Michigan.

Common Locust, False Acacia. **Robinia Pseudacacia, L.**
Ann Arbor; Mackinac, in cultivation.

Goat's Rue, Catgut. **Tephrosia Virginiana, Pers.**
S. W. (Wright); Livingston Co., (Miss Clark).

Milk-Vetch. **Astragalus Canadensis, L.**
Ann Arbor; Belle river, (Miss Clark); S. W. (Wright).

Tick Trefoil. **Desmodium nudiflorum, DC.**
S. Mich., (Wright).

Tick Trefoil. **Desmodium acuminatum, DC.**
S. Mich., (Wright).

Tick Trefoil. **Desmodium pauciflorum, DC.**
Mont Lake, (Miss Clark).

Tick Trefoil. **Desmodium rotundifolium, DC.**
S. Mich. (Wright).

Tick Trefoil. "Desmodium canescens, DC.?"
S. W. (Wright).

Tick Trefoil. Desmodium cuspidatum, Torr. & Gray. (*D. bracteosum*—*W.*)
S. Mich. (Wright).

Tick Trefoil. Desmodium laevigatum, DC.
S. Mich. (Wright).

Tick Trefoil. Desmodium Dillenii, Darlingt. (*D. Marylandicum*—*W.*)
S. W. [Wright); Mont L. (Miss Clark).

Tick Trefoil. Desmodium paniculatum, DC.
S. Mich. (Wright).

Tick Trefoil. Desmodium strictum, DC.
S. Mich. (Wright).

Bush Trefoil. Desmodium Canadense, DC.
Ann Arbor; Mont Lake, (Miss Clark).

Tick Trefoil. Desmodium sessilifolium, Torr. & Gray.
S. Mich., (Univ. Herb).

Tick Trefoil. Desmodium rigidum, DC.
Ann Arbor; S. W. (Univ. Herb).

Tick Trefoil. Desmodium ciliare, DC.
S. Mich., (Wright).

Tick Trefoil. Desmodium Marilandicum, Boott. (*D. obtusum*—*W.*)
S. Mich., (Wright).

Bush Clover. Lespedeza violacea, Pers.
S. W. (Wright); Ann Arbor, (Miss Clark).
var. angustifolia. (*L. reticulata.*—*W.*)
S. W. (Wright).

Slender Lespedeza. Lespedeza repens, Torr. & Gray. ("*L prostrata?*"—*W.*)
S. Mich., (Wright); Ann Arbor, (Miss Clark).

Bush Clover. Lespedeza Stuvei, Nutt.
S. Mich. (Univ. Herb).

Bush Clover. Lespedeza hirta, Ell. (*L. polystachia*—*W.*)
S. W. (Wright).

Bush Clover. Lespedeza capitata, Michx.
S. W. (Wright); Mont Lake, (Miss Clark).
var. angustifolia. (*L. angustifolia*—*W.*)
S. W. (Wright).

Vetch, Tare. Vicia Cracca, L.
S. Mich. (Wright).

Vetch. Vicia Caroliniana, Walt.
Ann Arbor, common.

Vetch. **Vicia Americana, Muhl.**
Ann Arbor; W. Mich. (Miss Clark).

Beach Pea. **Lathyrus maritimus, Bigelow.**
Pt. au Sable, Saginaw B., 17 June; shore of L. Huron, common; Lit. St. Martin's I.; S. W. (Univ. Herb.); Sand dunes of Emmet Co.

Vetchling. **Lathyrus venosus, Muhl.**
Ann Arbor.

Pale Vetchling. **Lathyrus ochroleucus, Hook.**
Ann Arbor; Pte au Chapeau, Saginaw B., 18 June. Among the settlers this species is called *Indian Pea.*

Marsh Vetchling. **Lathyrus palustris, L.**
Ann Arbor; Ft. Gratiot; Bay City, common; Psaganin, 26 June; Drummond's I., 26 July; Branch L., Antrim Co.

var. myrtifolius. (*L. myrtifolius—W.*)
Ft. Gratiot; Alpena Co., 6 July; Lit. St. Martin's I.; S. Mich. (Wright).

Kidney Bean. **Phaseolus diversifolius, Pers.**
S. Mich. (Wright).

Ground-nut. **Apios tuberosa, Moench.**
S. Mich. (Wright).

Hog Pea-nut. **Amphicarpaea monoica, Nutt.**
S. W. (Wright).

Wild Indigo. **Baptisia tinctoria, R. Br.**
Ann Arbor.

Baptisia leucantha, Torr. & Gray. (*B. alba.—W.*)
Calhoun County.

Baptisia leucophæa, Nutt.
S. Mich., (Torr. & Gr.)

Red-bud. **Cercis Canadensis, L.**
Ann Arbor.

Wild Senna. **Cassia Marilandica, L.**
Ann Arbor; S. W. (Wright).

Kentucky Coffee-tree. **Gymnocladus Canadensis, Lam.**
Ann Arbor.

Three-thorned Acacia, Honey Locust. **Gleditschia triacanthos, L.**
Ann Arbor.

ROSACEÆ.

Wild Yellow Plum, Red Plum. **Prunus Americana, Marshall.**
Ann Arbor; Pt. au Chene, L. Mich.

Sand Cherry. **Prunus pumila, L.** (*P. depressa—W.*)
Sand Point, Saginaw B.; shore of L. Huron to Drummond's I., (at Middle I. 5 feet high, branching diffusely from the base); Gros Cap, L. Mich.; very abundant on the sand dunes of Emmet Co.; and southward along the shore of L. Mich. The fruit is a black, medium sized cherry; flavor much like the choke-cherry, less astringent, but more bitter.

Wild Red Cherry. **Prunus Pennsylvanica, L.**

False Presqu' Isle; Drummond's I., very common; Sugar I.; Emmet Co.; S. E. (Wright). Small tree rarely exceeding 15 ft. in height.

Choke Cherry. **Prunus Virginiana, L.** (*P. obovata—W.*)

Ann Arbor; Sand Point, Saginaw B.; False Presqu' Isle; shore of L. Mich., Emmet and Antrim counties, abundant; N. shore of L. Mich.

Wild Black Cherry. **Prunus serotina, Ehrhart.**

Ann Arbor; Presqu' Isle; Emmet Co. Frequently attains the size of "a fine large tree." The largest specimens seen occur in Shiawassee Co., where it is an abundant forest tree.

Nine-Bark. **Spiræa opulifolia, L.**

Ann Arbor; Thunder B. Is., 7 July; Lit. St. Martin's I.; Drummond's I., common; Elk Rapids, Antrim Co., common; its clusters of white flowers, or red winged pods, making it one of the most attractive shrubs of the forest.

Common Meadow-sweet. **Spiræa salicifolia, L.**

Ann Arbor; Alpena Co., 6 July; Drummond's I.; Bruce Mine, Ca.; S. Mich. (Wright). Less common than the last.

Hardhack, Steeple-bush. **Spiræa tomentosa, L.**

S. W. (Wright); Mont Lake, (Miss Clark).

Queen of the Prairie. **Spiræa lobata, Murr.**

S. Mich. (Wright).

Bowman's Root. **Gillenia trifoliata, Moench.**

S. Mich. (Univ. Herb).

Common Agrimony. **Agrimonia Eupatoria, L.**

Ann Arbor; Ft. Gratiot; Pt. au Chene, L. Mich., 18 Aug.

Small-Flowering Agrimony. **Agrimonia parviflora, Ait.**

Detroit, (Miss Clark).

Canadian Burnet. **Sanguisorba Canadensis, L.**

S. Michigan, (Wright).

Avens. **Geum album, Gmelin.**

Shore of Saginaw B., 26 June; Pt. au Chene, L. Mich., 18 Aug.

Geum Virginianum, L.

Ann Arbor; Ft. Gratiot.

Large-leaved Avens. **Geum macrophyllum, Willd.**

Ft. Gratiot? (Austin.)

Geum strictum, Ait.

Ann Arbor; Ft. Gratiot; Bois Blanc I.; Ottawa, Iosco Co.

Water or Purple Avens. **Geum rivale, L.**

Ann Arbor; Mackinaw; Lit. St. Martin's I.

Barren Strawberry. **Waldsteinia fragarioides, Tratt.**

Livingston Co., 14 May; S. Mich. (Univ. Herb.)

Cinquefoil. **Potentilla Norvegica, L.**

Grass Island, Thunder Bay, 3 July; Drummond's I., common, 24 July; Ann Arbor, (Miss Clark).

Common Cinquefoil, Five-finger. **Potentilla Canadensis, L.**
Ann Arbor; Ft. Gratiot; Mouth of Saginaw R., 14 June; Mouth Sebawaing R., Tuscola Co.

Silvery Cinquefoil. **Potentilla argentea, L.**
Ann Arbor, (Miss Clark).

Potentilla arguta, Pursh.
Gros Cap, L. Mich., 18 Aug.; Ann Arbor, (Miss Clark).

Silver-weed. **Potentilla Anserina, L.**
Mouth of Sebawaing R., 14 June; shore of L. Huron, very common; Drummond's I.; Bruce Mine, Ca.; S. W. (Wright).

Shrubby Cinquefoil. **Potentilla fruticosa, L.**
Ann Arbor; Thunder Bay Is.; Drummond's I.; common on sandy and gravelly shores as well as near marshes, sometimes 4½ ft. in hight.

Marsh Five-finger. **Potentilla palustris, Scop.** (*P Comarum—W.*)
Ft. Gratiot; Sault Ste Marie, 28 July; Traverse City; S. E. (Wright).

Strawberry. **Fragaria Virginiana, Ehrhart.**
Ann Arbor; S. shore of Saginaw B., 13 June; Drummond's I.; Traverse City. More common than the next, except northward.

Strawberry. **Fragaria vesca, L.**
Ann Arbor; Middle I., L. Huron; Huron Co.; Mackinac.

Dalibarda. **Dalibarda repens, L.** [*D. fragaroides (violaeoides)—W.*]
Ann Arbor, (Miss Clark).

Purple Flowering Raspberry. **Rubus odoratus, L.**
Ft. Gratiot; Thunder Bay I., 7 July; Presqu' Isle, abundant, 12 July; Gros Cap, L. Mich.

White Flowering Raspberry. **Rubus Nutkanus, Mocino.**
Thunder Bay Is., 7 July; Presqu' Isle, 12 July. Earlier out of blossom than the last.

Dwarf Raspberry **Rubus triflorus, Richardson.** [*R. saxatilis (var. Canadensis)—W.*]
Ann Arbor; Lit. St. Martin's I., very abundant, trailing stems long and slender, covering the ground in shade of forests.

Wild Red Raspberry. **Rubus strigosus, Michx.**
Middle I., L. Huron; Thunder Bay Is., abundant; Bois Blanc I.; Sugar I., very abundant and very prolific; Emmet Co.; Mont Lake, (Miss Clark). Very common especially where the ground has been burnover. The fruit is largely manufactured into "raspberry jam" which is sent to all parts of the United States and to the W. Indies.

Black Raspberry, Thimbleberry. **Rubus occidentalis, L.**
Ann Arbor.

Common or High Blackberry. **Rubus villosus, Ait.**
Ann Arbor; Middle I., L. Huron; Drummond's I.; Emmet, Antrim and Grand Traverse counties, abundant.

var frondosus. (*R. frondosus.—W.*)
Traverse City; S. E. (Wright).

Low Blackberry, Dewberry. **Rubus Canadensis, L.**
Sand Pt., Saginaw B.; Saut St. Marie; Ann Arbor, (Miss Clark). Less common than the *R. villosus*; S. E. (Wright).

Running Swamp Blackberry. **Rubus hispidus, L.**
Squaw Pt , Thunder B.; Mont Lake, (Miss Clark); S. Mich. (Univ. Herb).

Low-bush Blackberry. **Rubus trivialis, Michx.**
S. Mich. (Wright); Mont Lake, (Miss Clark). Identification questionable.

Sand Blackberry. **Rubus cuneifolius, Pursh.**
(S.) Mich. (Miss Clark).

Climbing or Prairie Rose. **Rosa setigera, Michx.**
Jackson Co.; Gross Isle, (Miss Clark).

Swamp Rose. **Rosa Carolina, L.**
St. Joseph's I., 27 July; Ann Arbor.

Dwarf Wild-Rose. **Rosa lucida, Ehrhart.**
Ann Arbor; Drummond's I.; Sault Ste Marie.
var. parviflora, (Ehrhart). (*R. parviflora*—*W.*)
Sand Pt., Saginaw B.; S. Mich. (Wright).

Early Wild-Rose. **Rosa blanda, Ait.**
Ft. Gratiot; S. shore of Saginaw B.; Drummond's I.; St. Joseph's I., abundant, often forming the principal part of the shrubbery on high, rocky soil, or along gravelly beaches; Mackinac; Emmet Co.; Traverse City. The most frequent representative of this genus.

Sweet-Brier. **Rosa rubiginosa, L.**
Ann Arbor; Mackinac, abundant, 19 July.

Scarlet-fruited Thorn. **Cratægus coccinea, L.**
Ann Arbor; Stone I., Saginaw B., 16 June.

Black or Pear Thorn. **Cratægus tomentosa, L.**
Ann Arbor; Stone I., Saginaw B., 16 June.
var. pyrifolia.
Saut St. Marie; Ann Arbor, (Miss Clark).
var. punctata. (*C. punctata.*—*W.*)
Ann Arbor; Stone I., Saginaw B.
var mollis.
Ann Arbor.

Cockspur Thorn. **Cratægus Crus-galli, L.**
Ann Arbor.

Crab-Apple. **Pyrus coronaria, L.**
Ann Arbor.

Choke-berry. **Pyrus arbutifolia, L.**
Ann Arbor; Ft. Gratiot; Saut St. Marie.
var. melanocarpa. (*P. melanocarpa.*—*W.*)
Ann Arbor.

American Mountain-Ash. **Pyrus Americana, DC.**
St. Joseph I.

June berry. Shad-bush, Service-berry. **Amelanchier Canadensis, Torr. & Gr.**
Ann Arbor; St. Joseph's I.; Northport; Pt. au Chene, L. Mich.
var. Botryapium, (*A. Botryapium.*—*W.*)
Mackinac.
var. oblongifolia.
S. Mich. (Univ. Herb).

Medlar-bush. var. rotundifolia. (*A. ovalis.—W.*)
S. Mich. (Wright).
var. alnifolia.
Presqu' Isle.
var. oligocarpa, (*A sanguinea—W.*)
S. Mich. (Wright).

LYTHRACEÆ.

Ammannia. Ammannia humilis, Michx.
S. Mich. (Univ. Herb.)

Loosestrife. Lythrum alatum, Pursh.
S. Mich. (Univ. Herb.)

Spiked Loosestrife. Lythrum Salicaria, L.
S. Mich. (Wright).

Swamp Loosestrife. Nesæa verticillata, H. B. K. (*Decodon verticillatum —W.*)
S. Mich. (Wright). Gross Isle, (Miss Clark).

ONAGRACEÆ.

Great Willow-Herb. Epilobium angustifolium, L.
Ft. Gratiot; Alcona Co., 1 July; Thunder Bay, common; Drummond's I., common; Bruce Mine, Ca., common, a single specimen was found with white flowers; L. Sup.; Pt. au Chene, L. Mich. A very common and conspicuous herb, northward, especially where the ground has been burned over or cleared for settlement.

Epilobium palustre, L. Var. lineare. (*E. lineare—W.*)
Saut St. Marie.

Epilobium molle, Torr.
S. Mich. (Wright).

Epilobium coloratum, Muhl.
Ann Arbor; Ft. Gratiot; Middle I., L. Huron, 8 July; Saut St. Marie, common; Pt. au Chene, L. Mich., 19 Aug.; Traverse City.

Common Evening-Primrose. Œnothera biennis, L.
Ann Arbor; Thunder Bay Is., 3 July; Sugar I., common; Mackinac; Green R., Emmet Co.; S. W. (Wright).
var. muricata, (*Œ. muricata.—W.*)
S. W. (Wright).

Sundrops. Œnothera fruticosa, L.
Ann Arbor.

Œnothera pumila, L.
S. Mich. (Wright).

Gaura. Gaura biennis, L.
S. Mich. (Wright).

Seed-box. Ludwigia alternifolia, L.
S. W. (Wright).

False Loosestrife. Ludwigia polycarpa, Short & Peter.
Swamps, Michigan, (Dr. Pitcher).

Water Purslane. **Ludwigia palustris, Ell.**
S. Mich. (Univ. Herb.)

Enchanter's Nightshade. **Circæa Lutetiana, L.**
Ft. Gratiot; Pt. au Chene, L. Mich., 13 Aug.; Pine Lake, Emmet Co., 23 Aug. Found in moist, cold woodlands, not common.

Circæa alpina, L.
Ann Arbor; Ft. Gratiot; St. Joseph's I., 2 Aug.

Water Milfoil. **Myriophyllum verticillatum, L.**
S. Mich. (Wright).

Mare's-tail. **Hippuris vulgaris, L.**
S. Mich. (Wright).

GROSSULACEÆ.

Wild Gooseberry. **Ribes Cynosbati, L.**
Ann Arbor; Stone I., Saginaw B.; Drummond's I.

Smooth Wild Gooseberry. **Ribes hirtellum. Michx.**
Mackinac; Ann Arbor, (Miss Clark); Sitting Rabbit.

Smooth Wild Gooseberry. **Ribes rotundifolium, Michx.** (*R. triflorum.—W.*)
St. Joseph's I.; Fitting Rabbit; S. Mich. (Wright). The last two species of gooseberry were seen at Sitting Rabbit growing within three feet of each other, in a beach composed of fragments of limestone, very prolific. Though the former species is generally cultivated, the latter is preferable, the fruit being larger, with a pleasant tart in place of the flat sweetness of the former, and the branches less thorny. Its branches are spreading or procumbent; those of the former erect and rigid. By this difference they are easily distinguished at a distance.

Swamp Gooseberry. **Ribes lacustre, Poir.**
Drummond's I.; Sitting Rabbit; Grand Traverse Co.

Fetid Currant. **Ribes prostratum, L'Her.**
St. Joseph's I.

Wild Black Currant. **Ribes floridum, L'Her.**
Stone I., Saginaw B.; St. Joseph's I.; S. Mich. (Wright).

Red Currant. **Ribes rubrum, L.**
Ann Arbor.

CUCURBITACEÆ.

Wild Balsam-apple. **Echinocystis lobata, Torr. & Gr.** (*Mormordica echinata—W*)
S. Mich. (Wright).

CRASSULACEÆ.

Ditch Stonecrop. **Penthorum sedoides, L.**
Ann Arbor.

SAXIFRAGACEÆ.

Swamp Saxifrage **Saxifraga Pennsylvanica, L.**
Ann Arbor.

Common Alum-root. **Heuchera Americana, L.**
Ann Arbor.

Mitre-wort, Bishop's Cap. **Mitella diphylla, L.**
Ann Arbor.

Mitre-wort, Bishop's Cap. **Mitella nuda, L.** (*M. cordifolia.*—W.)
Pt. aux Barques, L. Huron, 21 June; Drummond's I.; Pittsfield, (Miss Clark).

False Mitre-wort. **Tiarella cordifolia, L.**
Ft. Gratiot; S. shore of Saginaw Bay; Bear Creek, Emmet Co., very abundant; Branch Lake, Antrim Co., 30 Aug.

Golden Saxifrage. **Chrysosplenium Americanum, Schwein.**
S. W. (Wright).

HAMAMELACEÆ.

Witch-Hazel. **Hamamelis Virginica.**
Ann Arbor; Mackinac; Traverse City; S. W. (Wright).

UMBELLIFERÆ.

Marsh Penny-wort. **Hydrocotyle Americana, L.**
Saut St. Marie, 30 July; Ann Arbor, (Miss Clark).

Marsh Penny-wort. **Hydrocotyle umbellata, L.**
S. W. (Wright).

Sanicle, Black Snakeroot. **Sanicula Canadensis, L.**
Ann Arbor.

Sanicle, Black Snakeroot. **Sanicula Marilandica, L.**
Ft. Gratiot; shore of Saginaw B., common; Drummond's I.; Pt. au Chene, L. Michigan; S. Mich. (Wright).

Rattlesnake-Master, Button Snakeroot. **Eryngium yuccæfolium, Michx.** (*E. aquaticum*—*W.*)
S. W. (Wright).

Polytænia Nuttallii, DC.
S. Mich. (Wright).

Cow Parsnip. **Heracleum lanatum, Michx.**
Ann Arbor; Stone I., Saginaw B., 16 June; Port Hope, Huron Co., abundant and very large; St. Helena I., L. Mich.

Common Parsnip. **Pastinaca sativa, L.**
Ann Arbor; Bois Blanc I.

Cowbane. **Archemora rigida, DC. var. ambigua,** (*A. ambigua*—*W*)
S. Mich. (Wright).

Archangelica. **Achangelica hirsuta, Torr. & Gr.** (*Angelica triquinata.*—W.)
Emmet Co.; S. W. (Wright); Ann Arbor, (Miss Clark).

Great Angelica. **Archangelica atropurpurea, Hoffm.** (*Angelica atropurpurea.*—W.)
Ann Arbor.

Meadow Parsnip. Thaspium barbinode, Nutt.
S. W. (Wright).

Meadow Parsnip. Thaspium aureum, Nutt.
Ann Arbor, rather common; S. shore of Saginaw Bay, common; Drummond's I.
var. apterum. (*Zizia aurea.*—W.)
S. Mich. (Wright).

Meadow Parsnip. Thaspium trifoliatum, Gray, var. apterum, Torr. & Gr. (*Zizia cordata.*—W.)
S. W. (Wright).

Alexanders. Zizia integerrima, DC.
Ann Arbor; Ft. Gratiot; Pt. au Chene, L. Mich.; Mackinac, (Miss Clark).

Spotted Cowbane Musquash-root. Cicuta maculata, L.
S. Mich. (Wright).

Cicuta bulbifera, L.
Ann Arbor; Grand Traverse Co.; Port Huron, (Miss Clark).

Water-Parsnip. Sium lineare, Michx.
S. Mich., (Univ. Herb).

Water-Parsnip. Sium angustifolium, L.
S. Mich. (Univ. Herb). *Sium latifolium* of Wright's Catalogue is probably one of these species.

Honewort. Cryptotænia Canadensis, DC.
Ann Arbor.

Smoother Sweet Cicely. Osmorhiza longistylis, DC.
Ann Arbor; Charity Is., 27 June; Pt au Chene, L. Mich.

Hairy Sweet Cicely. Osmorhiza brevistylis, DC.
Ft. Gratiot; shore of Saginaw Bay, 26 June; Pt. au Chene, L. Mich.; Ann Arbor. The prevailing species.

Poison Hemlock. Conium maculatum, L.
Mackinac, common.

Harbinger of Spring. Erigenia bulbosa, Nutt.
Ann Arbor, (Miss Clark).

ARALIACEÆ.

Spikenard. Aralia racemosa, L.
Ann Arbor; Sugar I., 31 July; Mackinac, (Miss Clark). Not common.

Bristly Sarsaparilla, Wild Elder. Aralia hispida, Michx.
Sturgeon Pt., L. Huron, 30 June, common; Drummond's I.; Pt. au Chene. L. Mich.; Emmet Co.; Port Huron, (Miss Clark).

Wild Sarsaparilla Aralia nudicaulis, L.
Ann Arbor; Pt. au Sable, Sag. Bay, 16 June; Drummond's I.; St. Joseph's I. Very common.

Ginseng. Aralia quinquefolia, Gray. (*Panax quinquefolium.*—W.)
S. W. (Wright); Saut St. Marie, and Ann Arbor, (Miss Clark).

Dwarf Ginseng, Ground-nut. Aralia trifolia, Gray. (*Panax trifolium.*—W.)
Ann Arbor.

CORNACEÆ.

Dwarf Cornel, Bunch-berry. **Cornus Canadensis, L.**
Ft. Gratiot; S. shore of Saginaw B., 18 June; Drummond's I., common; Sugar I.; St. Helena I.; Emmet Co., common; Leelanaw Co.; Pittsfield, (Miss Clark). Very common and widely diffused, northward.

Flowering Dogwood. **Cornus florida, L.**
S. Mich. (Wright).

Round-leaved Cornel. **Cornus circinata, L'Her.**
False Presvu' Isle, L. Huron, 11 July; S. Mich. (Wright).

Silky Cornel, Kinnikinnik. **Cornus sericea, L.**
Ann Arbor.

Red-osier Dogwood. **Cornus stolonifera, Michx.**
Stone I., Saginaw B., 16 June; Sand dunes of Ottawa Co., 30 Aug.; Ann Arbor.

Panicled Cornel. **Cornus paniculata, L'Her.**
Ann Arbor; Stone I., Saginaw Bay., 16 June; Bear Creek, Emmet Co.

Alternate-leaved Cornel. **Cornus alternifolia, L.**
Ann Arbor; Ft. Gratiot; Little Traverse Bay.

Pepperidge, Tupelo. **Nyssa multiflora, Wang.**
Ann Arbor; Bloomfield, Oakland Co.

CAPRIFOLIACEÆ.

Twin flower. **Linnæa borealis, Gronov.**
Pt. au Chapeau, Saginaw Bay, 18 June; shores of Lakes Huron and Michigan, very abundant.

Wolf-berry. **Symphoricarpus occidentalis, R. Br.**
Fort Gratiot, (Austin).

Snowberry. **Symphoricarpus racemosus, Michx.**
Pt. au Chapeau, Sag. Bay, 18 June; Alpena Co.

Yellow Honeysuckle. **Lonicera flava, Sims.**
Ann Arbor, (Miss Clark).

Small Honeysuckle. **Lonicera parviflora, Lam.**
Drummond's I., common.
var. Douglassii.
Ann Arbor; Pt. aux Barques, L. Huron, 19 June; Drummond's I.

Hairy Honeysuckle. **Lonicera hirsuta, Eaton**
Charity Is., Saginaw Bay, 27 June; Drummond's I., common; Pt. au Chene, L. Mich.

Fly Honeysuckle. **Lonicera ciliata, Muhl.** (*Hylosteum ciliatum.*—W.)
Sugar Island.

Bush Honey-Suckle. **Diervilla trifida, Mœnch.** (*D. Canadensis.*—W.)
Ann Arbor; Ft. Gratiot; Pt. au Barques, L. Huron, 19 June; shore of L. Huron, very common; St. Helena I.; Emmet, Antrim and Leelanaw counties, very common; Sugar I., abundant.

Fever-wort. **Triosteum perfoliatum, L.**
Ann Arbor; Ft. Gratiot.

Common Elder. **Sambucus Canadensis, L.**
Ann Arbor; Sanilac Co.

Red-berried Elder. **Sambucus pubens, Michx.** (*S. pubescens.—W.*)
Ann Arbor; Bois Blanc I.; Drummond's I.; Pt. au Chene, L. Mich.; Traverse City. More common northward than the last.

Sweet Viburnum. **Viburnum Lentago, L.**
Ann Arbor; Ft. Gratiot.

Downy Arrow-wood, Dock-Mackie. **Viburnum pubescens, Pursh.**
Ann Arbor.

Maple-leaved Arrow-wood. **Viburnum acerifolium, L.**
Ann Arbor; Ft. Gratiot; S. shore of Saginaw Bay, 28 June; Mission Pt., Grand Traverse Co.

Cranberry-tree. **Viburnum Opulus, L.** (*V. oxycoccus.—W.*)
Ann Arbor; Ft. Gratiot; shore of Saginaw Bay; St. Joseph's I.; Branch Lake, Antrim Co., abundant along the marshy margin of the river.

RUBIACEÆ.

Cleavers, Goose-Grass. **Galium Aparine, L.**
Saut St. Marie; S. W. (Wright); Ann Arbor, (Miss Clark).

Rough Bed-straw **Galium asprellum, Michx.**
Saut St. Marie, 29 July, growing rankly in the thickets near the river. One specimen measured 5 ft. 5 in. in hight, climbing and leaning on shrubs; Ann Arbor.

Galium concinnum, Torr. & Gr.
Ann Arbor.

Small Bed-straw. **Galium trifidum, L.**
Ann Arbor; S. shore of Saginaw Bay; Saut St. Marie, 29 July.
var. tinctorium, (*G. tinctorium.—W.*)
S. shore of Saginaw Bay, common; S. Mich., (Wright).
var. latifolium, (*G. obtusum.—W.*)
S. Mich., (Wright).

Sweet scented Bedstraw. **Galium triflorum, Michx.**
Willow river, shore of Sag. Bay, 20 June, common; Bruce Mine, Ca., 27 July; St. Helena I.; Ann Arbor, (Miss Clark). Very common throughout the northern portions of the State.

Galium pilosum, Ait.
Ann Arbor, (Miss Clark); S. Mich., (Univ. Herb.)

Wild Liquorice. **Galium circæzans, Michx.**
Ann Arbor; Ft. Gratiot.

Wild Liquorice. **Galium lanceolatum, Torr.**
S. Mich. (Wright).

Northern Bed-straw. **Galium boreale, L.**
Ann Arbor; Ft. Gratiot; S shore of Saginaw Bay.

Button-bush. **Cephalanthus occidentalis, L.**
Ann Arbor.

Partridge-berry. **Mitchella repens, L.**
Ft. Gratiot; Emmet Co., common; S. W. (Wright); Pittsfield, (Miss Clark).

Bluets. Oldenlandia purpurea.
Ann Arbor.
var. longifolia.
S. Mich., (Univ. Herb); Dover, (Miss Clark).
var. ciliolata, (*H. ciliolata.—W.*)
S. Michigan, (Wright).

VALERIANACEÆ.

Valerian. Valeriana sylvatica, Richards.
Ann Arbor.

Valerian. Valeriana edulis, Nutt.
Ann Arbor, (Miss Clark).

Corn Salad, Lamb-Lettuce. Fedia radiata, Michx.
Low grounds and moist fields, (Dr. Pitcher).

DIPSACEÆ.

Wild Teasel. Dipsacus sylvestris, Mill.
Ann Arbor.

COMPOSITÆ.

Iron-weed. Vernonia Noveboracensis, Willd.
S. Michigan, (Wright).

Iron-weed. Vernonia fasciculata, Michx.
S. W. (Univ. Herb).

Blazing Star. Liatris squarrosa, Willd.
Ann Arbor.

Button Snake-root. Liatris cylindracea, Michx.
S. Mich., (Wright).

Button Snake-root. Liatris scariosa, Willd.
Ann Arbor.

Gay-Feather. Liatris spicata, Willd.
S. W. (Wright).

Button Snake-root. Liatris pycnostachya, Michx.
Mont Lake, (Miss Clark). ?

Kuhnia. Kuhnia eupatorioides, L.
S. Mich. (Wright).

Joe-Pye Weed, Trumpet-Weed. Eupatorium purpureum, L.
Ann Arbor; Drummond's I; Bruce Mine, Ca., common; Pt. au Chene, L. Mich.; Mission Point; Saut St. Marie, common; Branch Lake, Antrim Co., abundant.
var. maculatum, (*E amoenum.—W.*)
S. Mich. (Wright).

Upland Boneset. Eupatorium sessilifolium, L.
S. Mich., (Wright).

Toroughwort, Boneset. Eupatorium perfoliatum, L.
Ann Arbor; Drummond's I., 22 July.

White Snake-root. **Eupatorium ageratoides, L.**
Ann Arbor.

Mist flower. **Conoclinium cœlestinum, DC.**
"Rich soil," (Gray).

Sweet Colts-foot. **Nardosmia palmata, Hook.**
Lake Huron, (Nuttall).

Colts-foot. **Tussilago Farfara, L.**
Saut St. Marie, (Whitney).

Corymbed Aster. **Aster corymbosus, Ait.**
S. Mich. (Wright).

Large Leaved Aster. **Aster macrophyllus, L.**
S. W. (Univ. Herb).

Silky Aster. **Aster sericeus, Vent.**
S. Michigan, (Wright).

Lax Leaved Aster. **Aster laxifolius, Nees.**
L. Huron, (Dr. Pitcher.)

Spreading Aster. **Aster patens, Ait.**
var. phlogifolius.
Ann Arbor, (Miss Clark).

Smooth Aster. **Aster lævis, L.**
Ann Arbor.
var. lævigatus.
Ann Arbor.

Azure Aster. **Aster azureus, Lindl.**
S. W. (Univ. Herb); Fort Gratiot, (Dr. Pitcher); Ann Arbor, (Miss Clark.)

Wavy Aster. **Aster undulatus, L.** ("*A. diversifolius?*"—*W.*)
S. Michigan, (Wright).

Heart Leaved Aster. **Aster cordifolius, L.** (*A. paniculatus.*—*W.*)
Ann Arbor; Drummond's I., common; Emmet Co.

Arrow Leaved Aster. **Aster sagittifolius, Willd.**
Ann Arbor; St. Joseph's I., 5 Aug.; S. W. (Univ. Herb).

Heath-like Aster. **Aster ericoides, L.**
Drummond's I., 9 Aug.; Ann Arbor, (Miss Clark).

Many Flowered Aster. **Aster multiflorus, Ait.**
Ann Arbor, 26 Sept., very common.

Tradescant's Aster. **Aster Tradescanti, L.**
Ann Arbor.

Dwarf Aster. **Aster miser, L., Ait.**
Bear Creek, Emmet Co., 24 Aug.; Ann Arbor, (Miss Clark).

Simple Aster. **Aster simplex, Willd.**
Leelanaw Co.

Thin Leaved Aster. **Aster tenuifolius, L.**
Emmet Co., 3 Sept.

Flesh colored Aster.

Aster carneus, Nees.
Pt. au Chene, L. Mich., 18 Aug.

Long Leaved Aster.

Aster longifolius, Lam. (*A. laxus.—W.*)
Ann Arbor.

New England Aster.

Aster Novæ-Angliæ, L.
Ann Arbor, (Miss Clark).

Sharp Leaved Aster.

Aster acuminatus, Michx.
"S. Michigan,?" (Wright).

Lofty Aster.

Aster præaltus, Poir. ("*A. salicifolius,?*"—*W.*)
S. Mich., (Wright). As this species is not embraced in Gray's Manual, Wright's determination may be regarded as exceedingly doubtful.

Sternutative Aster.

Aster ptarmicoides, Torr. & Gray.
Drummond's I., 10 Aug.; S. E. (Univ. Herb).

Horse-weed, Butter-weed.

Erigeron Canadense, L.
Ann Arbor; Drummond's I.; Saut St. Marie; Leelanaw Co.; Mackinac; Port Huron, (Miss Clark); S. W. (Wright). Very common everywhere.

Robin's Plantain.

Erigeron bellidifolium, Muhl.
Ann Arbor.

Fleabane.

Erigeron Philadelphicum, L.
Ann Arbor; Stone I., Saginaw B., 16 June; Drummond's I., 25 July.

Daisy Fleabane, Sweet Scabious.

Erigeron annuum, Pers. (*E. heterophyllum.—W.*)
Ann Arbor.

Daisy Fleabane.

Erigeron strigosum, Muhl.
Ann Arbor; S. shore of Sag. Bay, 21 June; Drummond's I., 9 Aug.

Golden-rod.

Solidago bicolor, L.
Gros Cap, L. Mich., 18 Aug. Rare.

var. concolor.
Pt. au Chene, L. Mich.; Drummond's I., common; Sugar I., very common; Alcona Co.

Solidago latifolia, L.
Ann Arbor.

Solidago cæsia, L. (*S. axillaris and flexicaulis.—W.*)
Ann Arbor, common; Bear Creek, Emmet Co., 24 Aug.; Traverse City; Northport. Common in the sandy soil of Emmet, Antrim, Grand Traverse and Leelanaw counties.

Solidago puberula, Nutt.
Presqu' Isle Co., 13 July, growing in a sandy beach; St. Joseph's I., 8 Aug., growing among other herbs and shrubs, in a gravelly soil, a few rods from the water.

Solidago stricta, Ait.
Drummond's I.

Solidago speciosa, Nutt.
Ann Arbor.

var. angustata.
Ann Arbor.

Solidago rigida, L.
Ann Arbor ; S. W. (Wright.)

Solidago Ohioensis, Riddell.
Drummond's I., 9 Aug.

Solidago Riddellii, Frank.
Emmet Co., 3 Sept.; S. W. (Univ. Herb).

Solidago Houghtonii, Torr & Gr.
Drummond's I., 25 July, plant sometimes 2 ft. in h pound corymb of 150 flowerheads.

Solidago patula, Muhl.
Ann Arbor.

Solidugo argūta, Ait.
Ann Arbor.
var. juncea, (*S. juncea—W.*)
S. Mich., (Wright).
var. scabrella.
Ann Arbor.

Solidago altissima, L.
Ann Arbor.

Solidago ulmifolia, Muhl.
S. W. (Univ. Herb).

Solidago nemoralis, Ait.
Ann Arbor; Drummond's I.; N. W. (Univ. Herb).

Solidago serotina, Ait.
S. Mich. (Wright).

Solidago Canadensis, L.
Ann Arbor; Drummond's I.. 25 July, common; Saut St. Marie; Emmet Co., comman; Northport, common; S. W. (Wright).

Solidago serotina, Ait.
S. Mich. (Wright).

Solidago lanceolata, L.
Drummond's I., 25 July; Pine Lake, 30 Aug.; S. W. (Wright).

Elecampane. Inula Helenium, L.
S. Michigan, (Wright).

Leaf-cup. Polymnia Canadensis, L.
S. Mich. (Wright).

Yellow Leaf-cup. Polymnia Uvedalia, L.
S. Mich., (Wright).

Rosin-weed, Compass-plant. Silphium laciniatum, L. (*S. gummiferum—W.*)
S. Mich., (Wright).

Prairie-dock. Silphium terebinthinaceum, L.
Ann Arbor ; S. W. (Wright).

Silphium integrifolium, Mich.
S. W. (Univ. Herb).

Cup-plant. Silphium perfoliatum, L.
S. Mich. (Wright).

Great Ragweed. Ambrosia trifida, L.
S. Mich. (Wright); Gross Isle, 3 Aug. (Miss Clark.)

Roman Wormwood, Hogweed, Bitter-weed. Ambrosia artemisiæfolia, L. (*A. elatior.—W.*)
Ann Arbor; very common.

Cocklebur. Clotbur. Xanthium strumarium, L.
S. Mich. (Wright).
var. echinatum.
S. Mich. (Univ. Herb).

Ox-eye. Heliopsis lævis, Pers.
Ann Arbor; S. W. (Wright).
var. scabra.
Ann Arbor, (Miss Clark).

Purple Cone-flower. Echinacea purpurea, Mœnch. (*Rudbeckia purparea —W.*)
S. W. (Univ. Herb).

Cone-flower. Rudbeckia laciniata, L.
Bear Creek, Emmet Co., 24 Aug.; S. W. (Wright); Northfield, (Miss Clark).

Cone-flower. Rudbeckia speciosa, Wender.
Ann Arbor.

Cone-flower. Rudbeckia fulgida, Ait.
Ann Arbor, (Miss Clark).

Cone-flower. Rudbeckia hirta, L.
Ann Arbor; Pt. au Chapeau, Saginaw Bay, 18 June; Drummond's I

Lepachys pinnata, Torr & Gr. (*Rudbeckia pinnata. —W.*)
S. Mich. (Wright).

Sunflower. Helianthus rigidus Desf.
Ann Arbor, 6 ft. in height.

Sunflower. Helianthus occidentalis, Riddell.
S. W. (Univ. Herb.)

Sunflower. Helianthus giganteus, L. (*H. giganteus and altissimus—W.*)
Ann Arbor.

Sunflower. Helianthus divaricatus, L.
Ann Arbor.

Sunflower. Helianthus hirsutus, Raf.
Ann Arbor, 26 Sept.; S. W. (Univ. Herb).

Sunflower. Helianthus strumosus, L.
S. Mich. (Wright).

Sunflower. Helianthus tracheliifolius, Wild.
S. Mich. (Wright).

Sunflower. Helianthus decapetalus, L. (*H. frondosus—W.*)
S. Michigan, (Wright).

Sunflower. Helianthus doronicoides, Lam.
Ann Arbor, (Miss Clark).

Actinomeris. Actinomeris squarrosa, Nutt.
S. Mich. (Wright).

Tickseed Sunflower. Coreopsis trichosperma, Michx.
S. Mich. (Wright).

Coreopsis aristosa, Michx.
S. Michigan, (Univ. Herb).

Tall Coreopsis. Coreopsis tripteris, L.
S. Mich. (Wright).

Coreopsis palmata, Nutt.
S. Mich. (Wright).

Coreopsis lanceolata, L.
L. Huron, 29 June; Dummond's I.; Traverse City.

Common Beggar-ticks. Bidens frondosa, L.
Ann Arbor; Northport, 11 Sept.

Swamp Beggar-ticks. Bidens connata, Muhl. (*B petiolata—W.*)
S. Mich. (Wright).

Bur-Marigold. Bidens cernua, L.
S. Mich. (Wright).

Bur-Marigold. Bidens chrysanthemoides, Michx.
Ann Arbor; Traverse City.

Water Marigold. Bidens Beckii, Torr.
S. Mich. (Wright).

Sneeze-weed. Helenium autumnale, L.
Ann Arbor.

Common May-weed. Maruta Cotula, DC.
Ann Arbor; Saut St. Marie; Emmet Co.; Northport. Very common everywhere.

Yarrow, Milfoil. Achillea Millefolium, L.
Ann Arbor; Stone I., Saginaw Bay; Drummond's I.; Saut St. Marie. Common.

Ox-eye Daisy. Leucanthemum vulgare, Lam.
Sand Pt., Saginaw Bay, 18 June; Bois Blanc I.

Cammon Tansy. Tanacetum vulgare, L.
Ann Arbor.

Tanacetum Huronense, Nutt.
Sand dunes of Emmet Co., common.

Canada Wormwood. Artemisia Canadensis, Michx.
Sand dunes of Ottawa Co., 30 Aug.; Sand dunes of Emmet Co., 25 Aug.; Drummond's I.

Western Mugwort. Artemisia Ludoviciana, Nutt.
var. gnaphalodes.
(Univ. Herb).

Everlasting. Gnaphalium decurrens, Ives.
Saut St. Marie, (Whitney).

Common Everlasting. Gnaphalium polycephalum, Michx.
Ann Arbor.

Low-Cudweed. Gnaphalium uliginosum, L.
Ann Arbor; Ft. Gratiot, (Miss Clark).

Pearly Everlasting. Antennaria margaritacea, R. Br.
Mackinac, 19 July.

Plantain-leaved Everlasting. Antennaria plantaginifolia, Hook. (*Gnaphalium plantagineum--W.*)
Ann Arbor; Stone I., Saginaw B.

Fireweed. Erechthites hieracifolia, Raf. *Senecio hieracifolius --W.*)
Leelanaw Co., 10 Sept.; S Mich. (Wright). Common, especially in the vicinity of recent clearings after the ground has been burned over, whence it receives its popular name.

Pale Indian Plantain. Cacalia atriplicifolia, L.
S. Mich. (Wright).

Tuberous Indian Plantain. Cacalia tuberosa, Mutt.
S. Mich. (Wright).

Cacalia suaveolens, L.
Lodi, (Miss Clark).

Common Groundsel. Senecio vulgaris, L.
* S. W. (Wright).

Golden Ragwort Squaw-weed. Senecio aureus, L.
The Cove, L. Huron, 1 July; S. Mich. (Univ. Herb).
var obovatus.
Ann Arbor.
var. Balsamitae. (*Senecio Balsamitae--W*)
Middle I., L. Huron, 1 July; Drummond's I., common; S. Michigan, (Wright). Throughout the northern shores of Lakes Huron and Mich., this variety is very common.

Common Thistle. Cirsium lanceolatum, Scop. (*Cnicus lanceolatus--W.*)
Ann Arbor, common; Mackinac.

Cirsium Pitcheri, Torr. & G. (*Cnicus Pitcheri--W.*)
Sand Pt. Saginaw B., 17 June; Emmet Co. Sandy shores.

Cirsium undulatum, Spreng.
Drummond's I.; 21 July.

Cirsium discolor, Spreng. (*Cnicus discolor--W.*)
S. Mich. (Wright).

Swamp Thistle. Cirsium muticum, Michx. (*Cnicus glutinosus--W.*)
Bruce Mine, Ca., 26 July; Drummond's I.; Emmet Co.; S. Michigan (Wright).

Pasture Thistle. Cirsium pumilum, Sprengr. (*Cnicus odoratus—W.*)
Drummond's I.? 21 July; S. W. (Wright).

Canada Thistle. Cirsium arvense, Scop.
Detroit, abundant; Ann Arbor, (Miss Clark).

Burdock. Lappa major, Gærtn. (*Arctium Lappa—W.*)
Ann Arbor; Huron Co.; S. W. (Wright); Mackinac.

Succory, Cichory. Cichorium Intybus, L.
Detroit, (Miss Clark).

Dwarf Dandelion. Krigia Virginica, Willd.
Psagrin. Bay Co., 26 June, rich, swampy soil; "Rockaway," (Miss Clark).

Cynthia. Cynthia Virginica, Don. (*Krigia amplexicaulis—W.*)
Ann Arbor, common; Ft. Gratiot; Pt. aux Gres, L. Huron.

Canada Hawkweed. Hieracium Canadense, Michx. (*H. Kalmii—W.*)
Saut St. Marie, 30 July; Sand dunes of Emmet Co., 21 Aug.

Rough Hawkweed. Hieracium scabrum, Mich. (*H. marianum—W.*)
Ann Arbor; Sand dunes of Emmet Co., 21 Aug.; S. Mich. (Wright); Port Huron, (Miss Clark).

Long-bearded Hawkweed. Hieracium longipilum, Torr. (*H. Scouleri—W.*)
Traverse City, 9 Sept.; S. W. (Wright).

Hairy Hawkweed. Hieracium Gronovii, L.
S. W. (Wright).

Rattlesnake-weed. Hieracium venosum, L.
Ann Arbor; Ft. Gratiot; Pigeon River, Sag. B., 18 June; Grand Traverse Bay.

Panicled Hawkweed. Hieracium paniculatum, L.
S. Mich. (Wright).

White Lettuce. Nabalus albus, Hook.
Ann Arbor; Pt. au Chene, L. Mich., 18 Aug.
var. Serpentaria, (*Prenanthes Serpentaria—W.*)
Ann Arbor; S. W. (Wright).

Tall White Lettuce. Nabalus altissimus, Hook.
Ann Arbor.

Nabalus racemosus, Hook. (*Prenanthes racemosa—W.*)
Shore of L. Mich., near Sitting Rabbit; S. W. (Wright).

Dandelion. Taraxacum Dens leonis, Desf. (*Leontodon Taraxacum—W.*)
Ann Arbor; Saginaw Bay; Saut St. Marie; S. W. (Wright).

Wild Lettuce. Lactuca elongata, Muhl.
S. W. (Wright).
var. sanguinea, Bigl. (*L. sanguinea—W.*)
S. W. (Wright).

False Blue Lettuce. **Mulgedium leucophaeum, DC.**
Ft. Gratiot; St. Joseph's I., common along St. Mary's River.

Spiny-leaved Sow-Thistle. **Sonchus asper, Vill.** (*S. oleraceus, var. asper—W.*)
S. E. (Wright).

LOBELIACEÆ.

Cardinal Flower. **Lobelia cardinalis, L.**
Ann Arbor; Bear Creek, Emmet Co., 24 Aug.

Great Lobelia. **Lobelia syphilitica, L.**
Ann Arbor; Branch Lake, Antrim Co., 30 Aug.

Lobelia spicata, Lam. (*L. Claytoniana—W*)
Ann Arbor; Ft Gratiot; mouth of Saginaw R., 24 June; Thunder Bay; Drummond's I.

Lobelia Kalmii, L.
Ann Arbor; Ft. Gratiot; Drummond's I., 25 July; S. W. (Wright).

CAMPANULACEÆ.

Harebell. **Campanula rotundifolia, L.**
Ann Arbor. Ft. Gratiot.
var. linifolia.
Ann Arbor; S. shore of Saginaw Bay, common. This is a very delicate and pretty species, occurring constantly and in every variety of situation. At Saginaw Bay it was in bloom in the middle of June, and was still abundantly in blossom Aug. 10th, at Grand Traverse Bay.

Marsh Bellflower. **Campanula aparinoides, Pursh.** (*C. erinoides—W.*)
St. Mary's River, 31 July; S. Mich. (Wright).

Tall Bellflower. **Campanula Americana, L.**
Ann Arbor.

Venus's Looking-glass. **Specularia perfoliata, A. DC.**
S. E. (Univ. Herb).

ERICACEÆ

Blue Tangle, Dangleberry. **Gaylussacia frondosa, Torr. & Gr.**
Ann Arbor.

Black Huckleberry. **Gaylussacia resinosa, Torr. & Gr.** (*Vaccinium resinosum—W*)
Ann Arbor; Grand Traverse Co.

Small Cranberry. **Vaccinium Oxycoccus, L.**
Ann Arbor.

Common American Cranberry. **Vaccinium macrocarpon, Ait.** (*Oxycoccus macrocarpus—W.*)
Ann Arbor; S. W. (Wright).

Dwarf Blueberry **Vaccinium Pennsylvanicum, Lam.**
Ann Arbor; St. Joseph's I., northern part. Abundant along the Canada Shore of St. Mary's R., producing abundance of fruit in the sparse soil of the hollows and crevices of metamorphic rocks; S. Mich. (Wright).

Canada Blueberry.

Vaccinium Canadense, Kalm.
Sitting rabbit; S. E. (wright).

Low Blueberry.

Vaccinum vacillans, Solander.
St. Joseph's I.; common in Emmet, Antrim, Grand Traverse and Leelanaw counties. S. Mich. (Univ. Herb).

Common Swamp Blueberry.

Vaccinium corymbosum, L.
Ann Arbor, (Miss Clark).

Creeping Snowberry.

Chiogenes hispidula, Torr. & Gr. (*Gaultheria hispidula—W.*)
S. Mich., (Wright).

Bearberry.

Arctostaphylos Uva-ursi, Spreng. (*Arbutus Uva-ursi—W.*)
Shores of L. Huron everywhere, very common; S. Mich. (Wright).

Trailing Arbutus, Ground Laurel.

Epigæa repens, L.
S. E. (Wright).

Aromatic Wintergreen.

Gaultheria procumbens, L.
Monroe Co.; Ottawa Co.; shores of L. Huron, very common; shore of L. Mich., Emmet, to Leelanaw Co., common; S. W. (Wright); Mont Lake, (Miss Clark).

Leather-leaf.

Cassandra calyculata, Don. (*Andromeda calyculata—W.*)
Livingston Co.; Shore of L. Mich., Emmet Co.; Drummond's I.; S. Mich. (Wright).

Wild Rosemary.

Andromeda polifolia, L.
S. Mich. (Wright); Ann Arbor, (Miss Clark). A shrub not distinguishable from this was seen at the mouth of Saginaw R., June 14, with corolla dark purple, awn wanting, pedicels dark brown 1¼ in. long from bracts.

Sheep Laurel, Lambkill.

Kalmia angustifolia, L.
Tawas City, 29 June, exquisitely beautiful and very abundant; Thunder Bay, common.

Swamp Laurel.

Kalmia glauca, Ait.
S. Mich. (Wright).

Labrador Tea.

Ledum latifolium, Ait.
Gros Cap., L. Mich.

Round-leaved Pyrola.

Pyrola rotundifolia, L.
Ft. Gratiot; Pt. au Pain Sucre, 19 June; St. Joseph's I., Little St. Martin's I.; Drummond's I., common; S. Mich. (Wright).
var asarifolia.
The Cove, L. Huron, 1 July.

Shin-leaf.

Pyrola elliptica, Nutt.
Ann Arbor; The Cove, L. Huron; Drummond's I.; St. Joseph I.; Grand Traverse Co.

Small Pyrola.

Pyrola chlorantha, Swartz.
Ft. Gratiot.

One-sided Pyrola

Pyrola secunda, L.
Ft. Gratiot; the Cove, L. Huron; St. Joseph's I.; Drummond's I.; S. W. (Wright).

One-flowered Pyrola. **Moneses uniflora, Gray.**
Ft. Gratiot; Little St. Martin's I.; 17 July, sweet scented.

Princes Pine, Pipsissewa. **Chimaphila umbellata, Nutt.** (*Pyrola umbellata—W.*)
Ft. Gratiot; L. Huron, Alcona Co.; L. Sup.; S. Mich. (Wright).

Pine drops. **Pterospora Andromedea, Nutt.**
Sitting Rabbit, 17 Aug.

Indian Pipe, Corpse-Plant. **Monotropa uniflora, L.**
Ann Arbor; Ft. Gratiot; Sitting rabbit.

Pine Sap, False Beech Drops. **Monotropa Hypopitys, L.**
Ft. Gratiot, (Austin).

AQUIFOLIACEÆ.

Black Alder, Winterbo.ry. **Ilex verticillata, Gray.** (*Prinos verticillatus—W.*)
S. W. (Wright); Ann Arbor, (Miss Clark).

Mountain Holly. **Nemopanthes Canadensis, DC.**
S. Mich. (Wright).

PLANTAGINACEÆ.

Common Plantain. **Plantago major, L.**
Ann Arbor; Saut Ste Marie; Mackinac.

Plantago cordata, Lam.
Tuscola Co.; S. Mich. (Wright).

Ribgrass, Ripplegrass, English Plantain. **Plantago lanceolata, L.**
Ann Arbor.

PRIMULACEÆ.

Bird's eye Primrose. **Primula farinosa, L.**
Drmmond's I.

Primula Mistassinica, Michx.
S. E. (Univ. Herb).

Chick-Wintergreen. **Trientalis Americana, Pursh.**
Ft. Gratiot; Pt. aux Barques, Sag. B., 21 June; St. Joseph's I.; Ann Arbor, (Miss Clark).

Loosestrife. **Lysimachia stricta, Ait.**
Ft. Gratiot; Saut St. Marie, 23 July; S. Mich. (Wright).

Lysimachia quadrifolia, L.
Ann Arbor.

Lysimachia ciliata, L.
Ann Arbor; Ft. Gratiot.

Lysimachia lanceolata, Walt. var. hybrida. (*L. hybrida—W.*)
S. W. (Wright).

Lysimachia longifolia, Pursh. (*L. revoluta—W.*)
Ann Arbor; Ft. Gratiot.

Tufted Loosestrife. **Naumburgia thyrsiflora, Reich.** (*L. Capitata—W.*)
Ann Arbor; Drummond's I., common in swampy soil; Sturgeon Pt., 30 June.

Common Pimpernel. **Anagallis arvensis, L.**
Ann Arbor, (Miss Clark).

Water Pimpernel, Brookweed. **Samolus Valerandi, L.**
Lodi, (Miss Clark).
var. Americanus.
N. W. (Univ. Herb.)

LENTIBULACEÆ.

Greater Bladderwort. **Utricularia vulgaris, L.** (*U. macrorhiza—W.*)
S. Mich. (Wright); Cape Ipperwash, C. W. (Austin).

Smaller Bladderwort. **Utricularia minor, L.** (*U. gibba—W.*)
Ann Arbor.

Utricularia intermedia, Hayne.
Ann Arbor.

Purple Bladderwort. **Utricularia purpurea, Walt.**
S. Mich. (Wright).

Horned Bladderwort. **Utricularia cornuta, Michx.**
Pt. au Chene, L. Mich., 18 Aug.; S. Mich. (Univ. Herb).

OROBANCHACEÆ.

Squaw root, Cancer-root. **Conopholis Americana, Wallroth.** (*Orobanche Americana—W.*)
Ann Arbor; Ft. Gratiot; S. W. (Wright).

One-flowered Cancer-root. **Aphyllon uniflorum, Torr. & Gr.** (*Orobanche uniflora—W.*)
S. E. (Wright).

SCROPHULARIACEÆ.

Common Mullein. **Verbascum Thapsus, L.**
False Presqu' Isle, L. Huron, 11 July; Grass Lake; Ann Arbor, common; Ft. Gratiot.

Moth Mullein. **Verbascum Blattaria, L.**
S. Mich. (Wright).

Wild Toad-Flax. **Linaria Canadensis, Spreng.**
S. shore Saginaw B., 17 June.

Toad-Flax, Butter-and-eggs, Ramsted. **Linaria vulgaris, Mill.**
Ann Arbor.

Figwort. **Scrophularia nodosa, L.** (*S. Marilandica and lanceolata—W.*)
Ann Arbor; S. W. (Wright).

Collinsia. **Collinsia verna, Nutt.**
Ann Arbor; S. W. (Wright); N. E. (Univ. Herb).

Turtle-head, Snake-head. **Chelone glabra, L.**
Ann Arbor; "Nebis" R., La Croix, Emmet Co.; S. W. (Wright). Its leaves vary from ½ in. to 1½ inches in diameter.

Beard-tongue, Penstemon. **Pentstemon pubescens, Solander.**
Ann Arbor, common.

Monkey-Flower. **Mimulus ringens, L.**
Ann Arbor; S. W. (Wright).

Monkey-Flower. **Mimulus alatus, Ait.**
S. W. (Wright).

Monkey-Flower. **Mimulus Jamesii, Torr.**
Mackinac, 17 July, abundant near the cool spring at the base of "Robinson's Folly;" St. Helena I., Straits of Mackinac, 20 Aug., abundant in wet, rich, low marshes. The plant is not always "*smooth*," being sometimes pubescent on the calyx, peduncles and lower side of the leaves. Both at Mackinac and St. Helena I. it was in company with *Veronica Americana, Schweinitz.*

Hedge Hyssop. **Gratiola Virginiana, L.**
S. Mich. (Univ. Herb).

False Pimpernel. **Ilysanthes gratioloides, Benth.** (*Lindernia attenuata and dilatata—W.*)
S. Mich. (Wright); Port Huron, (Miss Clark).

Synthyris. **Synthyris Houghtoniana, Benth.**
High prairies and hills, S. Mich. (Wright).

American Brooklime. **Veronica Americana, Schweinitz.** (*V. Beccabunga—W.*)
Ann Arbor; the Cove, L. Huron, 16 July; Mackinac. Common.

Culver's-root, Culver's Physic. **Veronica Virginica, L.**
Ann Arbor.

Water Speedwell. **Veronica Anagallis. L.**
Ann Arbor; S. W. (Wright).

Marsh Speedwell. **Veronica scutellata, L.**
Ann Arbor; Ft. Gratiot.

Common Speedwell. **Veronica officinalis, L.**
Ann Arbor.

Alpine Speedwell. **Veronica alpina, L.**
Saut St. Marie, 28 May.

Thyme-leaved Speedwell, Paul's Betony. **Veronica serpyllifolia, L.**
Ann Arbor; Bruce Mine, Ca., 26 July.

Neckweed, Purslane Speedwell. **Veronica peregrina, L.**
Ann Arbor, common.

Corn Speedwell. **Veronica arvensis, L.**
Ann Arbor, common.

Blue-hearts. **Buchnera Americana, L.**
S. W. (Wright); Mont Lake (Miss Clark).

Purple Gerardia. **Gerardia purpurea, L.**
S. Mich. (Wright); Mackinac, (Whitney).

Gerardia aspera, Dougl.
Sitting rabbit, 17 Aug., common; Pt. au Chene, L. Mich., abundant in sandy marshes.

Slender Gerardia. **Gerardia tenuifolia, Vahl.**
Ann Arbor.

Downy False-Foxglove. **Gerardia flava, L. partly.**
S. Mich. (Wright).

Smooth False-Foxglove. **Gerardia quercifolia, Pursh,** (*G. glauca—W.*)
S. W. (Wright); Mont Lake, Livingston Co., (Miss Clark).

Gerardia pedicularia L.
Traverse City, 9 Sept.; Ann Arbor.

Gerardia auriculata, Michx.
S. W. (Wright).

Scarlet Painted-cup. **Castilleia coccinea, Spreng.** (*Euchroma coccinea—W.*)
Ann Arbor, common; Ft. Gratiot; Mouth Saginaw R., a variety with yellow bracts instead of scarlet, 13 June; Mackinac; Drummond's I; Saut St. Marie.

Lousewort, Wood Betony. **Pedicularis Canadensis, L.**
Ann Arbor, common; False Presqu' Isle, L. Huron; Sugar I.; S. W. (Wright).

Pedicularis lanceolata, Michx (*P. pallida—W.*)
Ann Arbar; S. W. (Wright).

Cow-wheat. **Melampyrum Americanum, Michx.**
Ft. Gratiot; False Presqu' Isle, L. Huron, 11 July; Drummond's I., very common; L. Sup.; Mont Lake, (Miss Clark).

ACANTHACEÆ.

Water Willow. **Dianthera Americana, L.**
Ann Arbor.

Dipteracanthus ciliosus, Nees.
S. Mich. (Wright).

Dipteracanthus strepens, Nees. (*Ruellia strepens—W.*)
S. Mich. (Wright).

VERBENACEÆ.

Vervain. **Verbena angustifolia, Michx.**
S. Mich. (Univ. Herb).

Blue Vervain. **Verbena hastata, L.**
Bay City, 12 June; Bruce Mine, Ca., 25 July; Mackinac; Ann Arbor.

Nettle-leaved *or* White Vervain. **Verbena urticifolia, L.**
Ann Arbor.

Lopseed. Phryma Leptostachya, L.
Pt. au Chene, 18 Aug.; S. Mich. (Wright).

LABIATÆ.

Germander, Wood Sage. Teucrium Canadense, L.
Ann Arbor; S. W. (Wright).

Peppermint. Mentha Piperita, L.
Ann Arbor; S. W. (Wright).

Wild Mint. Mentha Canadensis, L. (*M. borealis*—*W.*)
Ann Arbor; Bruce Mine, Ca., 26 July; Drummond's I.; Sugar I. Common about the shores of L. Huron.

Bugleweed. Lycopus Virginicus, L.
Ann Arbor; Bruce Mine, Ca., 25 July; Pte Ste Ignace, common, corolla has five almost equal lobes, probably owing to the large upper lobe being 2-cleft, and often a small additional calyx tooth between the bases of the regular ones.

Water Horehound. Lycopus Europæus, L.
S. Mich. (Wright).
var. sinuatus.
Drummond's I., 22 July.

Hyssop. Hyssopus officinalis, L.
S. W. (Univ. Herb).

Mountain Mint, Basil. Pycnanthemum lanceolatum, Pursh. (*P. Virginicum*—*W.*)
S. Mich. (Wright).

Mountain Mint. Basil. Pycnanthemum linifolium, Pursh.
Ann Arbor, moist woods and exsiccated swamps.

Calaminth. Calamintha glabella, Benth. var. Nuttallii.
Drummond's I., 22 July, in crevices of limestone rocks, very common. This plant has a strong savor like the *American Pennyroyal*, for which it is often mistaken, especially by the settlers throughout the northern lake shores where the true *American Pennyroyal* has not, as yet, been found; S. E. (Wright).

Basil. Calamintha Clinopodium, Benth.
Ft. Gratiot.

American Pennyroyal. Hedeoma pulegioides, Pers.
S. Mich. (Wright).

Hedeoma hispida, Pursh. (?)
Middle I., L. Huron, 9 July.

Horse Balm, Rich-weed, Stone-Root. Collinsonia Canadensis, L.
Ann Arbor; S. W. (Wright); Elmwood, Detroit, (Miss Clark).

Oswego Tea, Bee Balm. Monarda didyma, L.
Ft. Gratiot, (Austin).

Wild Bergamot. Monarda fistulosa, L. (*M. allophylla*—*W.*)
Ann Arbor; Emmet Co., 22 Aug., common in sandy soil; S. Mich. (Wright).

Horse-mint. Monarda punctata, L.
S. Mich. (Wright).

Blephilia.

Blephilia ciliata, Raf.
Alpena Co. (?) 6 July, having four perfect stamens and two strongly awned teeth on the lower lip of the calyx; Ann Arbor; Saut St. Marie.

Blephilia hirsuta, Benth.
S. Mich. (Wright).

Giant Hyssop.

Lophanthus nepetoides, Benth. (*Hyssopus nepetoides—W.*)
S. W. (Wright); Grosse Isle, (Miss Clark).

Lophanthus scrophulariæfolius, Benth. (*Hyssopus scrophulariæfolius—W.*)
S. Mich. (Wright).

Cat-mint, Catnip.

Nepeta Cataria, L.
Ann Arbor; Drummond's I, 9 Aug.

False Dragon head.

Physostegia Virginiana, Benth. (*Dracocephalum Virginianum—W.*)
Ann Arbor; S. W. (Wright).

Heal-all, Self-heal.

Brunella vulgaris, L. (*Prunella vulgaris—W.*)
Ann Arbor, common; Alpena, 6 July; Drummond's I., a variety with white corolla; Saut St. Marie.

Skullcap.

Scutellaria versicolor, Nutt. (*S. cordifolia—W.*)
S. Mich. (Wright).

Scutellaria pilosa, Michx.
S. W. (Univ. Herb).

Scutellaria integrifolia, L.
S. Mich. (Univ. Herb).

Scutellaria parvula, Michx. (*S. ambigua—W.*)
S. Mich. (Wright).

Scutellaria galericulata, L.
Ann Arbor; Bay Co., common, 27 June; Port Huron, (Miss Clark).

Mad-dog Skull-cap.

Scutellaria lateriflora, L.
Sulphur I., (north of Drummond's,) 8 Aug., common; Ann Arbor.

Horehound.

Marrubium vulgare, L.
S. Mich. (Univ. Herb).

Hemp-nettle.

Galeopsis Tetrahit, L.
Mackinac, 19 July, common; Sugar I., very abundant, but it was noticed that the upper lip of the corolla is not entire but almost always with three or four teeth at its apex; and that the three lobes of the lower lip are similar, the middle one a little larger,—all oval; S. E. (Wright).

Red Hemp-nettle.

Galeopsis Ladanum, L
Ft. Gratiot; Saut St. Marie.

Hedge-nettle.

Stachys palustris, L. var. aspera, (*S. aspera—W.*)
S. W. (Wright); Ann Arbor, (Miss Clark).

Stachys hyssopifolia, Michx.
S. Mich. (Wright).

Motherwort. **Leonurus Cardiaca, L.**
Pine Lake, Emmet Co., 29 Aug.; Ann Arbor.

BORRAGINACEÆ.

Common Comfrey. **Symphytum officinale, L.**
Ann Arbor; Port Austin, Huron Co.

Corn Gromwell. **Lithospermum arvense, L.**
Ann Arbor.

Common Gromwell. **Lithospermum officinale, L.**
Mackinac, 16 July; S. Mich. (Wright).

Lithospermum latifolium, Michx.
S. Mich. (Univ. Herb).

Hairy Puccoon. **Lithospermum hirtum, Lehm.**
Sand Pt., Saginaw B., 17 June, abundant; Monroe Co., (Miss Clark); Ft. Gratiot.

Hoary Puccoon. **Lithospermum canescens, Lehm.** (*Batschia canescens—W.*)
Ann Arbor, common.

Early Forget-me-not. **Myosotis verna, Nutt.**
Ann Arbor; (Dr. Lord).

Stick-seed. **Echinospermum Lappula, Lehm.**
Ann Arbor; Mackinac, 18 July.

Hound's tongue. **Cynoglossum officinale, L.**
Ann Arbor; Ft. Gratiot.

Wild Comfrey. **Cynoglossum Virginicum, L.** (*C. amplexicaule—W.*)
Ft. Gratiot; Presqu' Isle, L. Huron; S. Mich. (Wright).

Beggar's Lice. **Cynoglossum Morrisoni, DC.**
Ann Arbor.

HYDROPHYLLACEÆ.

Waterleaf. **Hydrophyllum Virginicum, L.**
Ann Arbor.

Hydrophyllum Canadense, L.
S. Mich. (Wright).

Hairy Waterleaf. **Hydrophyllum appendiculatum, Michx.**
Ann Arbor; S. W. (Wright).

POLEMONIACEÆ.

Wild Sweet William. **Phlox maculata, L.**
Rich woods and riverbanks, (Gray).

Carolina Phlox. **Phlox Carolina, L.**
S. Mich. (Univ. Herb).

Hairy Phlox. **Phlox pilosa, L.**
Ann Arbor.

Divaricate Phlox. Phlox divaricata, L.
Ann Arbor.

Ground or Moss Pink. Phlox subulata, L.
S. Mich. (Univ. Herb).

CONVOLVULACEÆ.

Wild Potato-vine, Man-of-the-earth. Ipomœa pandurata, Meyer.
Ann Arbor.

Bindweed. Convolvulus arvensis, L.
Ann Arbor.

Hedge Bindweed. Calystegia sepium, R. Br.
Ft. Gratiot; S. W. (Wright); Gross Isle, (Miss Clark).
var. repens.
S. shore of Sagin. w B., 14 June.

Low Bindweed. Calystegia spithamæa, Pursh.
Stone I., Saginaw Bay, 16 June; S. E. (Wright).

Dodder. Cuscuta Gronovii, Willd. (*C. Americana—W.*)
S. W. (Wright).

Cuscuta glomerata, Choisy.
Moist prairies, (Gray).

SOLANACEÆ.

Bittersweet. Solanum Dulcamara, L.
Ann Arbor; Pine Lake, 29 Aug.

Common Nightshade. Solanum nigrum, L.
Ann Arbor; Sugar I., 31 July; S. Mich. (Wright).

Ground Cherry. Physalis pubescens, L. (*P. obscura—W.*)
Ann Arbor.

Ground Cherry. Physalis viscosa, L.
Ann Arbor; Drummond's I., 23 July.

Apple of Peru. Nicandra physaloides, Gaertn.
Ann Arbor.

Black Henbane. Hyoscyamus niger, L.
Ft. Gratiot; Mackinac, 19 July, abundant.

Stramonium. Datura Stramonium, L. (*D Tatula—W.*)
Ann Arbor.

Wild Tobacco. Nicotiana rustica, L.
Emmet Co., 25 Aug., cultivated by the Indians.

GENTIANACEÆ.

American Centaury. Sabbatia angularis, Pursh.
S. Mich. (Wright).

American Columbo. Frasera Carolinensis, Walt.
Jackson Co. and westward; Ann Arbor, (Miss Clark).

Spurred Gentian. **Halenia deflexa, Griseb.**
Middle I., L. Huron, 9 July; Drummond's I.; St. Helena I., common. S. E. (Univ. Herb).

Five-flowered Gentian. **Gentiana quinqueflora, L.**
Ann Arbor.
var. occidentalis.
Ann Arbor.

Fringed Gentian. **Gentiana crinita, Froel.**
Ann Arbor; Mackinac, (Whitney).

Smaller Fringed Gentian. **Gentiana detonsa, Fries.**
Ann Arbor; Drummond's I., 13 Aug.; Pt. au Chene, L. Mich.

Straw Colored Gentian. **Gentiana ochroleuca, Froel.**
Mont Lake, (Miss Clark).

Whitish Gentian. **Gentiana alba, Muhl.**
Ann Arbor; S. W. (Univ. Herb).

Closed Gentian. **Gentiana Andrewsii, Griseb.**
S. Mich. (Univ. Herb).

Soapwort Gentian. **Gentiana Saponaria, L.**
S. Mich. (Wright).
var. linearis.
Pt. au Chene, L. Mich., 19 Aug., sandy swamps.

Gentiana puberula, Michx.
Ann Arbor, (Miss Clark).

Screw-stem. **Bartonia tenella, Muhl.** (*Centaurella paniculata—W.*)
S. W. (Wright).

Buckbean. **Menyanthes trifoliata, L.**
Ann Arbor; S. Mich. (Wright).

APOCYNACEÆ.

Spreading Dogbane. **Apocynum androsæmifolium, L.**
Ann Arbor; The Cove, L. Huron, 1 July; St. Joseph's I.

Indian Hemp. **Apocynum cannabinum, L. var. glaberrimum.**
Ann Arbor; Ft. Gratiot.
var. pubescens, DC.
Ft. Gratiot; Charity I., 27 June.
var. hypericifolium. (*A. hypericifolium—W.*)
S. Mich. (Wright).

ASCLEPIADACEÆ.

Milkweed, Silkweed. **Asclepias Cornuti, Decaisne.** (*A. Syriaca—W.*)
Ann Arbor; Charity Is., 27 June; Sand dunes, Emmet Co.

Poke Milkweed. **Asclepias phytolaccoides, Pursh.**
Ann Arbor, very short pedicels except the terminal one; Ft. Gratiot.

Purple Milkweed. **Asclepias purpurascens, L.**
Ann Arbor.

Variegated Milkweed. **Asclepias variegata, L.**
Ann Arbor.

Four-leaved Milkweed. **Asclepias quadrifolia, Jacq.**
Ann Arbor.

Swamp Milkweed **Asclepias incarnata, L.**
Ann Arbor; Ft. Gratiot; Grand Traverse Co.; S. W. (Wright).

Butterfly-weed, Pleurisy-root. **Asclepias tuberosa, L.**
Ann Arbor; Ft. Gratiot.

Whorled Milkweed. **Asclepias verticillata, L.**
S. Mich. (Wright).

Green Milkweed. **Acerates viridiflora, Ell.** (*Asclepias lanceolata—W.*)
Ft. Gratiot; S. W. Mich. (Wright).

OLEACEÆ.

White Ash. **Fraxinus Americana, L.** (*F. acuminata—W.*)
Ann Arbor; Drummond's I.; Emmet Co. Common in the Southern Peninsula, but apparently less frequent northward.

Red Ash. **Fraxinus pubescens, L.**
Drummond's I.; S. Mich. (Wright). Comparatively rare.

Green Ash. **Fraxinus viridis, Michx. f.**
Ann Arbor.

Black Ash, Water Ash. **Fraxinus sambucifolia, Lam.**
Ann Arbor; Sugar I., common; Pine Lake; S. W. (Wright).

Blue Ash. **Fraxinus quadrangulata, Michx.**
S. Mich. (Univ. Herb). The wood of the Ash is highly esteemed for its strength and suppleness, especially the first and last species above. The White Ash is most common and most extensively used, its annual growths being least liable to separate into layers. It is much preferable for oars, being light as well as tough when seasoned. It is also extensively used by fishermen for hoops and staves, but for this the Black Ash is always preferred from the greater ease with which its layers are separated.

The Black Ash is a smaller tree, and is generally found in the vicinity of swamps or along streams. The value of its timber is increased by the rapidity of its growth. It is tougher and more elastic than the White Ash, but less durable upon exposure to the vicissitudes of moisture and dryness. North of the Straits of Mackinac this is the prevailing species. The Blue Ash is found only in the southern part of the State. Its timber is prized equally with that of the White Ash, for which it is substituted in many of its uses. The Red Ash is a smaller tree and furnishes less valuable timber.

ARISTOLOCHIACEÆ.

Wild Ginger. **Asarum Canadense, L.**
Ann Arbor.

PHYTOLACCACEÆ.

Poke, Scoke, Garget, Pigeon-berry. **Phytolacca decandra, L.**
S. Mich. (Wright).

CHENOPODIACEÆ.

Maple-leaved Goosefoot. **Chenopodium hybridum, L.**
Ann Arbor; Drummond's I.; Mackinac.

Lamb's Quarters, Pigweed. **Chenopodium album, L.**
Ann Arbor; St. Joseph's I.

Jerusalem Oak, Feather Geranium. **Chenopodium Botrys, L.**
Ann Arbor; S. W. (Wright.)

Mexican Tea. **Chenopodium ambrosioides, L.**
S. W. (Wright); Ann Arbor, (Miss Clark).

"Chenopodium rubrum, L."
S. Mich. (Wright). [Probably a synonym of one of the preceding species.]

Strawberry Blite **Blitum capitatum, L.**
Pt. aux Barques, L. Huron, 20 June; Drummond's I.; Emmet Co. Common about the shores of lakes Huron and Michigan.

AMARANTACEÆ.

Green Amaranth, Pigweed. **Amarantus hybridus, L.**
Ann Arbor.

Prince's Feather. **Amarantus hypochondriacus, L.**
Ann Arbor, (Miss Clark).

Pigweed. **Amarantus retroflexus, L.**
Ann Arbor.

Amarantus albus, L.
Ann Arbor, (Miss Clark).

Montelia tamariscina, Gray.
S. Mich. (Univ. Herb).

Acnida cannabina, L.
S. Mich. (Wright).

POLYGONACEÆ.

Prince's Feather. **Polygonum orientale, L.**
Ann Arbor.

Water Persicaria. **Polygonum amphibium, L.**
Ft. Gratiot; Saginaw Bay, 16 June.
var. aquaticum, L.
Ann Arbor; St. Mary's R. 1 Aug.; Mont Lake, (Miss Clark).

Polygonum nodosum, Pers. var. incarnatum.
Ann Arbor.

Polygonum Pennsylvanicum, L.
S. Mich. (Wright).

Lady's Thumb. **Polygonum Persicaria, L.**
Ann Arbor.

Smartweed. **Polygonum Hydropiper, L.**
Ann Arbor.

Wild Smartweed. **Polygonum acre, H. B. K.** *(P. punctatum—W.)*
Ann Arbor; Mackinac, (Miss Clark).

Mild Water-pepper. **Polygonum hydropiperoides, Michx.** (*P. mite—W.*)
S. Mich., (Wright).

Knotgrass, Goosegrass, Door-weed. **Polygonum aviculare, L.**
Ann Arbor; Bruce Mine, Ca., 26 July.
var. erectum, Roth.
Ann Arbor.

Slender Knotgrass. **Polygonum tenue, Michx.**
S. Mich. (Wright).

Jointweed. **Polygonum articulatum, L.**
Traverse City, 8 Sept., beginning to blossom, abundant.

Polygonum Virginianum, L.
Ann Arbor; S. W. (Wright).

Halberd-leaved Tear-thumb. **Polygonum arifolium, L.**
Gros Cap, L. Mich., 18 Aug.; S. Mich. (Wright).

Arrow-leaved Tear-thumb. **Polygonum sagittatum, L.**
Saut St. Marie 31 July; S. Mich. (Wright).

Black Bindweed. **Polygonum Convolvulus, L.**
S. W. (Wright).

Polygonum cilinode, Michx.
Drummond's I., common; Huron Co., 20 June; Saut St. Marie.

Climbing False Buckwheat. **Polygonum dumetorum, L.** (*P. scandens—W.*)
Ann Arbor; Mackinac, (Miss Clark).

Buckwheat. **Fagopyrum esculentum, Moench.** (*Polygonum Fagopyrum—W.*)
Ann Arbor.

Swamp Dock. **Rumex verticillatus, L.** (*R. Brittanica—W.*)
Ann Arbor; Islands of Thunder Bay, 7 July; S. W. (Wright).

Tall Dock. **Rumex altissimus, Wood.**
Saut St. Marie, 29 July.

Willow Dock. **Rumex salicifolius, Weinmann, Hook.**
Villa Cross, Emmet Co., 22 Aug.

Great Water Dock. **Rumex Hydrolapathum, Hudson, var. Americanum, Gray.**
Bruce Mine, Ca., 26 July; S. Mich. (Univ. Herb).

Bitter Dock. **Rumex obtusifolius, L.**
Saut St. Marie, 29 July; Ann Arbor, (Miss Clark).

Curled Dock. **Rumex crispus, L.**
Ann Arbor; Saut St. Marie, 29 July. Commonest species of dock.

Bloody-vined Dock. **Rumex sangiuneus, L.**
Ann Arbor, (Miss Clark).

Field or Horse Sorrel. **Rumex Acetosella, L.**
Ann Arbor; Ft. Gratiot; Saginaw B.; Drummond's I.; Mackinac. Common.

LAURACEÆ.

Sassafras. **Sassafras officinale, Nees.** (*Laurus Sassafras—W.*)
Ann Arbor.

Fever-bush, Spice-bush, Benjamin-bush, Wild Allspice. **Benzoin odoriferum, Nees.** (*Laurus Benzoin—W.*)
S. Mich. (Wright); Ann Arbor.

THYMELEACEÆ.

Leatherwood, Moose-wood. **Dirca palustris, L.**
Ann Arbor.

ELÆAGNACEÆ.

Shepherdia. **Shepherdia Canadensis, Nutt.**
Ann Arbor; Drummond's I., common about rocky shores.

SANTALACEÆ.

Bastard Toad-flax. **Comandra umbellata, Nutt.**
Ann Arbor; Ft. Gratiot; Sand Pt., Saginaw Bay, 18 June, common. *C. livida* occurs at Cove I., L. Huron.

SAURURACEÆ.

Lizard's Tail. **Saururus cernuus, L.**
S. Mich. (Wright).

EUPHORBIACEÆ.

Shore Spurge. **Euphorbia polygonifolia, L.**
S. Mich. (Wright); Fort Gratiot.

Spotted Spurge. **Euphorbia maculata, L.**
Ann Arbor; S. W. (Wright); Grosse Isle, (Miss Clark); Ft. Gratiot.

Flowering Spurge. **Euphorbia corollata, L.**
Ann Arbor.

Euphorbia commutata, Englm.
Ann Arbor.

Three-seeded Mercury. **Acalypha Virginica, L.**
Ann Arbor; S. Mich. (Wright).

URTICACEÆ.

Slippery or Red Elm. **Ulmus fulva, Michx.**
Ann Arbor; Emmet Co.; Branch Lake, Antrim Co. Less common than the next.

American or White Elm. **Ulmus Americana, L.**
Ann Arbor; Drummond's I.; Sugar I., large and common in the low lands of this island; Antrim Co.; Saut St. Marie, several fine native specimens stand east of the town near the river.

Corky White Elm. **Ulmus racemosa, Thomas.**
Ann Arbor, in a swamp about a mile south of the city.

The Elm prefers low grounds and rich soils. It is especially flourishing at the head of Branch Lake, Antrim Co., where the Red and the White Elm were found growing large and promiscuously together, the latter, however, excelling in size.

Sugarberry. Hackberry. **Celtis occidentalis, L.**
Grosse Isle, (Miss Clark).
var. crassifolia, (*C. crassifolia—W.*)
S. W. (Wright).

Tall wild Nettle. **Urtica gracilis, Ait.**
Ann Arbor; Saut St. Marie, 29 July.

Great Stinging Nettle. **Urtica dioica, L.**
Gros Cap, L. Mich., 18 Aug.; S. W. (Wright).

Wood Nettle. **Laportea Canadensis, Gaudich.** (*Urtica Canadensis —W.*)
Ann Arbor; Pt. au Chene, L. Mich., 18 Aug.; Pine Lake; Lodi, (Miss Clark).

Richweed, Clearweed. **Pilea pumila, Gray.** (*Urtica pumila—W.*)
S. W. (Wright); Ann Arbor, (Miss Clark).

Boehmeria cylindrica, Willd. (*Urtica capitata—W.*)
S. W. (Wright).

Hemp. **Cannabis sativa, L.**
Ann Arbor; S. W. (Wright); Mackinac, (Miss Clark).

Hop. **Humulus Lupulus, L.**
Ann Arbor.

Plane, Sycamore. **Platanus occidentalis, L.**
Ann Arbor.

JUGLANDACEÆ.

Butternut. **Juglans cinerea, L.**
Ann Arbor.

Black Walnut. **Juglans nigra, L.**
Ann Arbor. Generally throughout the southern part of the Peninsula.

Shell-bark or Shag-bark Hickory. **Carya alba, Nutt.**
Ann Arbor, &c., common.

Thick Shell-bark Hickory. **Carya sulcata, Nutt.**
Ann Arbor, &c., common.

Small-fruited Hickory. **Carya microcarpa, Nutt.**
Ann Arbor.

Pignut, Broom Hickory. **Carya glabra, Torr.**
Ann Arbor, common.

Bitter-nut, Swamp Hickory. **Carya amara, Nutt.**
Ann Arbor. This genus is abundantly represented in Southern Michigan.

CUPULIFERÆ.

Bur-Oak. **Quercus macrocarpa, Michx.**
Ann Arbor; S. W. (Wright). Common.

White Oak. **Quercus alba, L.**
Ann Arbor; Grand Traverse Bay. Common throughout the southern peninsula.

Swamp White Oak. **Quercus Prinus, L.**
var. discolor, Michx. (*Q. bicolor—W.*)
Ann Arbor.

Yellow Chestnut Oak. **Quercus Castanea, Willd.**
Ann Arbor.

Chinquapin or Dwarf Chestnut Oak. **Quercus prinoides, Willd.**
Ann Arbor. ?

Laurel or Shingle Oak. **Quercus imbricaria, Michx.**
Ann Arbor.

Quercitron or Black Oak, Yellow-barked Oak **Quercus tinctoria, Bartram.**
Ann Arbor.

Scarlet Oak. **Quercus coccinea, Wang.**
Ann Arbor; Traverse City.

Red Oak. **Quercus rubra, L.**
Drummond's I.; Sugar I., common; Emmet Co., in the northern part of this county attains the largest size seen, growing in sandy soil in the valleys of the sand dunes, and producing fruit in great abundance. One tree measured 8 feet in circumference, 3 feet above the surface. The lee side of these dunes is covered more or less with trees and shrubs of the general character of the flora of the region, imbedded in the sand sometimes half their height.

Swamp Spanish, or Pin Oak. **Quercus palustris, Du Roi.**
Ann Arbor. The oak family is abundantly represented in the southern peninsula, forming a great part of the forest timber. Still the immediate shore of L. Huron from Bay county to the Straits of Mackinac is apparently entirely destitute of oaks. Thence northward along St. Mary's river *Q. rubra* is the only species, which is quite common but never attains a large size. Southward this species is of rarer occurrence, but is replaced by other and more valuable species. At Traverse City *Q. alba* and *Q. coccinea* make their most northern appearance as far as observed. There they are about equally frequent, growing in a sandy soil, sometimes in patches surrounded by the pines and more or less dispersed amongst them, the former forming a large and shapely trunk, the latter being a smaller, but well proportioned tree. South of this the species multiply both in numbers and frequency of occurrence.

Chestnut. **Castanea vesca, L.**
Monroe Co. Not common in the southern peninsula.

American Beech. **Fagus ferruginea, Ait.** (*F. sylvatica—W.*)
Ann Arbor; Mackinac, common, but so reduced in size as to be of little value; Drummond's I.; shore of L. Mich., from the Straits of Mackinac to Northport, the beech here forming a very large and valuable part of the forest growth. Here were seen the largest and most perfect specimens. In the southern counties it is very common, and furnishes excellent fuel.

Wild Hazelnut. **Corylus Americana, Walt.**
Ann Arbor; Mackinac.

Beaked Hazelnut. **Corylus rostrata, Ait.**
St. Joseph's I.; Drummond's I.

Hornbean, Blue or Water Beech. Iron-wood. **Carpinus Americana, Michx.**
Ann Arbor.

Hop-Hornbean, Lever-wood. Iron-wood. **Ostrya Virginica, Willd.**
Ann Arbor; Drummond's I.; Emmet Co., common.

MYRICACEÆ.

Bayberry, Wax-Myrtle. **Myrica cerifera, L.**
S. Mich. (Univ. Herb.)

Sweet fern. **Comptonia asplenifolia, Ait.**
Alpena; Traverse City; Ottawa Co.; Oakland Co., (Miss Clark); S. Mich. (Wright).

BETULACEÆ.

Paper Birch, Canoe Birch. **Betula papyracea, Ait.**
Gravelly Pt., L. Huron; False Presqu' Isle, L. Huron; Drummond's I.; Sugar I.; Emmet Co. This is a very common tree throughout the lake shores, growing in the most unfavorable situations, but seldom forming a large trunk. It is apt to spring up as second growth where the forest has been destroyed by fire. There is a variety (?) known as "red birch" by the Indians and Half-Breeds, with pale reddish bark much more brittle than the ordinary.

Yellow Birch. **Betula excelsa, Ait.**
Ann Arbor; Pt. aux Barques, L. Huron; Pt. au Chene, L. Mich.; Emmet Co.; Drummond's I. Less common that the preceding, but often grows to a large tree. One specimen in Antrim Co. had a circumference of 11 ft. 4 in., four feet above the ground.

Cherry Birch, Sweet or Black Birch. **Betula lenta, L.**
Drummond's I., only seen at this place, but attains a monstrous size, one specimen measuring 10 feet in circumference. This is a valuable tree, and it is unfortunate that so little is found in the State.

Low Birch. **Betula pumila, L.** (*B. glandulosa—W.*)
S. E. (Wright).

Speckled or Hoary Alder. **Alnus incana, Willd.**
Shores of L. Huron; Drummond's I., common; Saut St. Marie, and along the banks of St. Mary's river, abundant.

Smooth Alder. **Alnus serrulata, Ait.**
Traverse City; S. Mich. (Wright).

SALICACEÆ.

Hoary Willow. **Salix candida, Willd.**
Ann Arbor; Drummond's I.; north shore of Little L. George, very abundant, growing in the shallow margin of the lake.

Low Bush Willow. **Salix humilis, Marshall.**
Drummond's I.

Glaucous Willow. **Salix discolor, Muhl.**
Ann Arbor; Drummond's I.; Pine Lake, Emmet Co., abundant along the margin of the lake, occasionally reaching the size of a small tree. One tree measured 6½ inches in diameter a foot from the surface.

Silky-headed Willow. **Salix eriocephala, Michx.**
Ann Arbor; Drummond's I.

Silky-leaved Willow. **Salix sericea, Marshall.**
Ann Arbor? Drummond's I.

Petioled Willow. **Salix petiolaris, Smith.**
Saut St. Marie; S. E. (Univ. Herb).

Heart-leaved Willow. **Salix cordata, Muhl.**
Grand Traverse Co.

Narrow-leaved Willow. **Salix angustata, Pursh.**
Ann Arbor.

Long-beaked Willow. **Salix rostrata, Richardson.**
Ann Arbor; St. Joseph's I.; S. E. (Univ. Herb). A common species.

Brittle Willow. **Salix fragilis, L.**
Ann Arbor.

Black Willow. **Salix nigra, Marshall.**
Ann Arbor.

Shining Willow. **Salix lucida, Muhl.**
Drummond's I.; S. E. (Univ. Herb).

Long-leaved Willow. **Salix longifolia, Muhl.**
S. Mich. (Univ. Herb.)

Stalk-fruited Willow. **Salix pedicellaris, Pursh.**
Drummond's I.; S. E. (Univ. Herb).

Weeping Willow. **Salix Babylonica, L.**
Cultivated in many places for ornament. Barely spontaneous.

American Aspen. **Populus tremuloides, Michx.**
Ann Arbor; Sand Pt., Saginaw B.; Drummond's I.; Sugar I., this and the next were growing abundant, large and promiscuously together; Antrim Co. This is one of the most common trees about the lake shores, seldom attaining, however, a large size. It often springs up abundantly with *Betula papyracea* in exposed situations where the forest growth has been prostrated by fire or tempest.

Large-toothed Aspen. **Populus grandidentata, Michx.**
Ann Arbor; Sand Pt., Saginaw Bay; Sugar I.; Antrim Co This is a rarer but more valuable tree than the last. They are often found in company, but this was never known to accept an exposed or unfavorable situation for the sake of such company; while the former often intrudes upon soil and situations chosen by the latter. This often forms a large tree with a naked, smooth and dingy yellow trunk expanding its brawny and conspicuous limbs at a height of 50 feet.

Downy-leaved Poplar. **Populus heterophylla, L.**
S. Mich. (Univ. Herb.)

Cotton wood, Necklace Poplar. **Populus monilifera, Ait.** (*P. Canadensis—W.*)
Ann Arbor.

Balsam Poplar, Tacamahac. **Populus balsamifera, L.**
Thunder Bay, L. Huron; L. Mich., common. This is a common tree in low lands about the borders of rivers and swamps. It is very common on Drummond's I., but it is of little economical value, its height seldom reaching 30 feet, while its usual size is about fifteen.

Balm of Gilead. **var. candicans.** (*P. candicans—W.*)
This is common in cultivation, but rare in a wild state. But a single specimen was seen which was a large tree standing near the lake shore a few miles north of Elk Rapids, Antrim C

Lombardy Poplar. **Populus dilatata, Ait.**
Ann Arbor, in cultivation, and at many other localities, sparingly naturalized. Probably the largest specimens of this tree in the State are on the site of the "Old British Fort" near the mouth of St. Mary's R., on Drummond's I.

CONIFERÆ.

Gray or Northern Scrub Pine.

Pinus Banksiana, Lambert.

Sand Pt. Saginaw Bay, and northward along the shore of L. Huron, not common.

Red Pine.

Pinus resinosa, Ait.

Pt. au Chapeau, Sag. B., northward along the shore of L. Huron, Drummond's I., and the shore of L. Mich., both on the Upper and Lower Peninsula, common. This is improperly called "Norway Pine" by the lumbermen.

Pitch Pine.

Pinus rigida, Miller.

S. Mich. (Wright).

White Pine.

Pinus Strobus, L.

Abundant in the valley of the Saginaw R. and its branches; shore of L. Huron; Drummond's I.; Sugar I., huge solitary specimens of the species were seen overtopping the surrounding forest, generally large but not abundant; Shore of L. Mich. to Traverse City; Ottawa Co., &c.

Of the Pines, the last is most valuable and most abundant. In the valley of the Saginaw river, within 21 miles of its mouth there are fifty steam-saw mills which are employed upon the logs of this species principally, and within the space of three miles there may be seen no less than 21 mills. These logs are "poled" down the river and its branches from the pine lands through which they flow. Also on the south shore of Saginaw B., and at Pigeon River, Pinnebog, Port Austin and Willow River, Huron Co., the lumber business is extensively carried on. Also northward, along the shore of L. Huron, wherever there are facilities for transporting the logs by means of the small streams, mills have been erected for the manufacture of lumber. Along the northern shore, the "Norway Pine" becomes frequent. At Elk Rapids and Traverse City, *P. resinosa* is more extensively sawed, which furnishes less valuable lumber for general purposes, though very suitable for flooring. This is generally found in sandy soil on level tracts forming "Pine plains," the trees having tall, straight and naked trunks, and presenting a beautiful orchard-like appearance. The shore of L. Michigan produces the white and red pine in about equal abundance, nowhere forming exclusive forests, but rather standing alone or in small clusters in the midst of surrounding Beeches, Maples and Hemlocks. *P. Banksiana* is a small irregular tree of little value, preferring the most exposed and barren situations.

Balsam Fir.

Abies balsamea, Marshall.

Shores and islands of L. Huron, very common; St. Mary's river and shore of L. Mich., everywhere common. This is the prevailing species of the genus; and next to the Arbor Vitæ, is the most frequent of the family *Coniferæ A. Canadensis* prefers higher land among the beeches, large poplars and birches, more inland; while this occupies the lower, more recent drift, nearer the lake. Often it is seen growing at the base of a ledge of rocks, while at the summit and further inland *A. Canadensis* grows large and abundant.

Small-fruited or Double Balsam Fir.

Abies Fraseri, Pursh.

N. W. (Univ. Herb).

Hemlock Spruce.

Abies Canadensis, Michx.

Shore of L. Huron, but not common, from Huron county northward; Drummond's I.; Emmet and Antrim counties. Nowhere is this species known to excel the large and majestic growth which it attains on the shores of L. Mich., in Antrim county. Indeed, with this exception, it was rarely met with during the entire season. It is found also in Ottawa, Kent and Allegan counties.

Black, or Double Spruce.

Abies nigra, Poir.

Whitmore Lake, Washtenaw Co.; The Cove, L. Huron; Drummond's I.; common; Sugar I.; shore of L. Mich., less common.

White, or Single Spruce. **Abies alba, Michx.**
Drummond's I., common; Sugar and St. Joseph's Is.; Shore of L. Mich., rare. The black spruce is more widely diffused over the State than the white, but the white predominates in the northern districts.

Larch, Tamarack, Hackmatack. **Larix Americana, Michx.** (*Pinus pendula—W.*)
Ann Arbor; False Presqu' Isle, L. Huron; Drummond's I.; Sugar I., abundant and very large, rising to the hight of 100 feet or more, with a circumference of six and a half feet, two feet above the ground. Swampy lands at the head of Branch Lake, Antrim Co., very large. This is a common tree in low, marshy land, and often mingles with the white cedar in the well known "cedar swamps."

Arbor Vitæ, White Cedar. **Thuja occidentalis, L.**
The most striking and hardy tree of the forest, growing in all situations, with its roots immersed in water, and on the most barren and inaccessible heights. In low and level tracts it often forms extensive "cedar swamps." It may always be seen about the shore of an inland lake or the margin of a river, its dense foliage or dry scraggy limbs projecting over the water. It is generally a tree about 25 ft. in height, but sometimes grows to a monstrous size on high lands where there is soil sufficient to sustain it. The largest specimens seen were growing in the sandy soil of Emmet Co., in higher situations than is usual for the tree. One specimen among others scattered through the forest composed mostly of beeches and hemlocks, had a diameter of 4 ft. 2 in. four feet from the ground. It is of slow growth, and requires centuries to attain such dimensions. Its wood has the greatest durability and is much used for fence posts, while its bark furnishes thatching for the wigwam of the Indian and the cabin of the settler; S. Michigan (Wright). On Drummond's Island were seen willows 25 to 30 years old, growing above the prostrate trunks of the white cedar, still remaining in a perfect state of preservation. In other cases it is equally sound beneath peat bogs, or buried 30 feet under "modified drift," where it must have lain for ages.

Juniper. **Juniperus communis, L**
Ann Arbor; Pt. aux Barques, L. Huron; False Presqu' Isle, L. Huron; Old Ft. Mackinac, Emmet Co. Common about the lake shores.

Red Cedar, Savin. **Juniperus Virginiana, L.**
Ann Arbor; Thunder Bay Is.; N. W. (Univ. Herb); S. Mich. (Wright).

var. humilis, Hook.
False Presqu' Isle, with trailing stems 25 ft. long; Sand dunes, Emmet Co., abundant.

American Yew Ground Hemlock. **Taxus baccata, L. var. Canadensis, Gray.**
Middle I., L. Huron, very abundant; Drummond's I., common; L. Mich. Common throughout the northern counties, especially in the shade of evergreens, a declining, one-sided shrub, having a luxuriant dark-green foliage and presenting a beautiful appearance where it covers the surface, but vieing with the White Cedar in forming a most persistent obstruction to the progress of the pedestrian.

ARACEÆ.

Indian Turnip. **Arisæma triphyllum, Torr.** (*Arum triphyllum—W.*)
Ann Arbor; Ft. Gratiot; Northport.

Green Dragon, Dragon-root. **Arisæma Dracontium, Schott.**
Ann Arbor.

Arrow Arum. **Peltandra Virginica, Raf.** (*Rensselaeria Virginica —W.*)
S. Mich. (Wright).

Water Arum. **Calla palustris, L.**
Ann Arbor; S. W. (Wright).

Skunk Cabbage. Symplocarpus fœtidus, Salisb. (*Ictodes fœtidus—W.)*
Ann Arbor; Northport.

Sweet Flag, Calamus. Acorus Calamus, L.
Quanecussee, Tuscola Co.; Bruce Mine, Ca.

TYPHACEÆ.

Cat-tail Flag. Typha latifolia, L.
Ann Arbor, common; Saginaw B., common; Saut St. Marie.

Bur-reed. Sparganium eurycarpum, n. sp. Englm.
Ann Arbor; Saginaw Bay, 14 June.

Bur-reed. Sparganium ramosum, Hudson.
S. Mich. (Wright); Ft. Gratiot.

Bur-reed. Sparganium simplex, Hudson. (*S. Americanum—W.*)
Ft. Gratiot; Saginaw Bay, 14 June; S. E. (Wright).

LEMNACEÆ.

Duckweed, Duck's-meat. Lemna trisulca, L.
S. E. (Wright).

Duckweed. Lemna minor, L.
S. W. (Univ. Herb); Northfield, Washtenaw Co., (Miss Clark).

Duckweed. Lemna polyrhiza, L.
S. Mich. (Univ. Herb).

NAIADACEÆ.

Pondweed. Potamogeton pectinatus, L.
S. Mich. (Wright).

Pondweed. Potamogeton pauciflorus, Pursh.
S. E. (Univ. Herb).

Pondweed. Potamogeton perfoliatus, L.
S. Mich., (Wright).

Pondweed. Potamogeton prælongus, Wulf.
Saginaw Bay, 16 June; (Univ. Herb).

Pondweed. Potamogeton lucens, L.
S. Mich., (Wright).
var. ? fluitans.
S. E. (Univ. Herb)

Pondweed. Patamogeton natans, L.
S. Mich., (Wright).

Pondweed. Potamogeton heterophyllus, Schreber.
S. Mich. (Wright).

ALISMACEÆ.

Arrow-grass. Triglochin palustre, L.
S. W. (Wright).

Arrow-grass. Triglochin maritimum, L.
S. Mich., (Wright).
var. elatum.
False Presqu' Isle, L. Huron, 11 July, common; Drummond's I.; Ann Arbor, (Miss Clark).

Scheuchzeria. Scheuchzeria palustris, L.
S. W. (Wright).

Water Plantain. Alisma Plantago, L. var. Americanum, Gray. (*A. Plantago—W.*)
Ann Arbor; S. W. (Wright).

Arrow-head. Sagittaria variabilis, Engelm.
"Psaganing," Bay Co., 26 June; Ann Arbor.
var. diversifolia.
S. Mich., (Univ. Herb).
var. angustifolia.
Elk Rapids, Antrim Co.

Arrow-head. Sagittaria heterophylla, Pursh.
Elk Rapids, Antrim Co.

Arrow-head. Sagittaria pusilla, Nutt.
S. W. (Univ. Herb).

HYDROCHARIDACEÆ.

Waterweed. Anacharis Canadensis, Planchon.
S. Mich. (Univ. Herb).

Tape grass, Eel grass. Vallisneria spiralis, L.
Ann Arbor.

ORCHIDACEÆ.

Showy Orchis. Orchis spectabilis, L.
Ann Arbor. Near the light-house at the mouth of Saginaw river is a variety with light purple lip, interruptedly streaked and mottled with dark purple. In bloom, 14 June.

Naked-gland Orchis. Gymnadenia tridentata, Lindl. (*Habenaria tridentata—W.*)
S. W. (Wright).

Large Round-leaved Orchis. Platanthera orbiculata, Lindl. (*Habenaria orbiculata—W.*)
Ft. Gratiot; False Presqu' Isle, L. Huron, 11 July; Drummond's I.; Saut Ste Marie, (Miss Clark). Rare.

Smaller two-leaved Orchis. Platanthera Hookeri, Lindl.
S. E. (Univ. Herb).

Bracted Green Orchis. Platanthera bracteata, Torr. (*Habenaria bracteata—W.*)
Ann Arbor; Emmet Co.

Northern Green Orchis. Platanthera hyperborea, Lindl. (*Habenaria huronensis—W.*)
Ann Arbor; Squaw Pt., Thunder Bay, 6 July; Northport; S. W. (Wright). Common.

Northern White Orchis. Platanthera dilatata, Lindl.
Drummond's I., 22 July; S. E. (Univ. Herb).

Yellowish Orchis Platanthera flava, Gray. (*Habenaria herbiola—W.*)
Ann Arbor; S. W. (Wright).

Yellow Fringed Orchis. Platanthera ciliaris, Lindl. (*Habenaria ciliaris—W.*)
Ann Arbor.

White Fringed Orchis. Platanthera blepharigllottis, Lindl.
S. Mich., (Univ. Herb).

Western Orchis. Platanthera leucophæa, Nutt.
Ann Arbor.

Ragged Orchis. Platanthera lacera, Gray. (*Habenaria psycodes, partly—W.*)
Ann Arbor.

Small Purple Fringed-Orchis. Platanthera psycodes, Gray. (*Habenaria psycodes, partly, H grandiflora and fimbriata—W.)*
Ft. Gratiot; Drummond's I., 22 July; S. W. (Wright).

Large Purple Fringed-Orchis. Platanthera fimbriata, Lindl.
Milford, Oakland Co.; Ann Arbor, (Miss Clark). *P. obtusata* occurs at Cove I., L. Huron, (Austin).

Rattlesnake Plantain. Goodyera repens, R. Br.
Antrim Co., 3 Sept., common in the shade of woods.

Rattlesnake Plantain. Goodyera pubescens, R. Br.
Ann Arbor.

Ladies' Tresses. Spiranthes gracilis, Big.
S. W. (Wright).

Ladies' Tresses. Spiranthes latifolia, Torr. in Lindl.
Drummond's I., common; S. E. (Univ. Herb).

Ladies' Tresses. Spiranthes cernua, Richard.
Ann Arbor.

Arethusa. Arethusa bulbosa, L.
S. Mich. (Wright).

Pogonia. Pogonia ophioglossoides, Nutt.
S. W. (Wright).

Pogonia. Pogonia pendula, Lindl. (*Triphora pendula—W.*)
S. W. (Wright).

Calopogon. Calopogon pulchellus, R. Br.
Ann Arbor; Mouth Saginaw R., 24 June; S. Mich. (Wright).

Calypso. Calypso borealis, Salisb.
Forty-mile point, Presqu' Isle Co.

Crane-Fly Orchis. Tipularia discolor, Nutt.
N. Mich. (Dr. Cooley).

Adder's-Mouth. Microstylis ophioglossoides, Nutt.
S. W. (Wright).

Adder's-Mouth. Microstylis monophyllos, Lindl.
Ann Arbor, (Miss Clark).

Twayblade. Liparis liliifolia, Richard. (*Malaxis liliifolia—W.*)
S. W. (Wright).

Coral-root. Corallorhiza innata, R. Br. (*C. verna—W.*)
S. E. (Wright).

Coral-root. Corallorhiza multiflora, Nutt.
Pt. aux Barques, Huron Co., 20 June; St. Martin's I., 17 July; S. W. (Wright).

Coral-root. Corallorhiza odontorhiza, Nutt.
Rich woods, (Gray.)

Coral-root. Corallorhiza Macræi, Gray.
Mackinac, (C. G. Loring, Jr., and Whitney).

Putty-root, Adam-and-Eve. Aplectrum hyemale, Nutt.
S. E. (Univ. Herb).

Larger Yellow Lady's Slipper. Cypripedium pubescens, Willd.
Ann Arbor; Stone I., Saginaw B., 16 June; Drummond's I.

Small White Lady's Slipper. Cypripedium parviflorum, Salisb.
Ann Arbor.

Showy Lady's Slipper. Cypripedium candidum, Muhl.
Ann Arbor.

Smaller Yellow Lady's Slipper. Cypripedium spectabile, Swartz.
Ann Arbor; Tawas Bay, 28 June.

Stemless Lady's Slipper. Cypripedium acaule, Ait.
Ann Arbor; Grand Rapids, (Miss Clark). *C. arietinum* occurs at Cape Ipperwash, C. W., a few miles from Port Huron.

AMARYLLIDACEÆ.

Star-grass. Hypoxys erecta, L.
Ann Arbor, common; Ft. Gratiot; shores of Sag. B., common.

HÆMODORACEÆ.

Colic-root, Star-grass. Aletris farinosa, L.
S. E. (Wright).

IRIDACEÆ.

Larger Blue Flag. Iris versicolor, L.
Ann Arbor; Ft. Gratiot; Saginaw Bay, common; Mackinac. Common all over the Southern Peninsula.

Lake Dwarf Iris. Iris lacustris, Nutt.
Bois Blanc I.; Mackinac; Drummond's I.; Old Fort Mackinac.

Blue-Eyed Grass. Sisyrinchium Burmudiana, L.
Ann Arbor, very common; Ft. Gratiot; shores of Saginaw Bay 14 June, common.

var. anceps, (*S. anceps—W.*)
S. W. (Wright).

DIOSCOREACEÆ.

Wild Yam-root. Dioscorea villosa, L.
S. W. (Wright); Ann Arbor.

SMILACEÆ.

Common Greenbrier. Smilax rotundifolia, L.
S. Mich. (Wright).

Smilax hispida, Muhl.
Ann Arbor.

Carrion Flower. Smilax herbacea, L.
Ann Arbor.
var. pulverulenta, (*S. peduncularis—W.*)
S. Mich. (Wright).

Smilax tamnifolia, Michx.
Ann Arbor.

Nodding Trillium, Wake Robin. Trillium cernuum, L.
S. Mich., (Wright).

Purple Trillium, Birthroot. Trillium erectum, L.
Ann Arbor.
var. album.
Ann Arbor.

Large White Trillium. Trillium grandiflorum, Salisb.
Ann Arbor; Drummond's I. A variety occurs at Ann Arbor with flowers tetramerous throughout.

Painted Trillium. Trillium erythrocarpum, Michx.
S. Mich. (Wright).

Indian Cucumber-root. Medeola Virginica, L. (*Gyromia Virginica—W.*)
Alcona Co., 1 July; S. Mich. (Wright).

LILIACEÆ.

Smaller Solomon's Seal. Polygonatum biflorum, Ell.
Ann Arbor.

Great Solomon's Seal. Polygonatum giganteum, Dietrich. (*Convallaria multiflora—W.*)
Ann Arbor.

False Spikenard. Smilacina racemosa, Desf.
Ann Arbor; Drummond's I.

Smilacina stellata, Desf.
Ann Arbor; Ft. Gratiot; Sand dunes, Ottawa Co., 30 Aug., but 3-seeded ! ; Huron Co.

Smilacina trifolia, Desf.
S. Mich. (Univ. Herb).

Smilacina bifolia, Ker
Ann Arbor; Ft. Gratiot. Common everywhere.

Clintonia borealis, Raf.
Common in shady, moist woods throughout the northern counties of the peninsula.

Wild Leek. Allium tricoccum, Ait.
St. Martin's I., 17 July; S. W. (Wright).

Wild Onion. Allium cernuum, Roth.
S. W. (Wright); Ann Arbor, (Miss Clark).

Wild Meadow Garlic. Allium Canadense, Kalm.
Ann Arbor; S. shore of Saginaw Bay, 14 June; S. Mich. (Wright).

Wild Orange-red Lily. Lilium Philadelphicum, L.
Ann Arbor; Ft. Gratiot; Stone I., Saginaw B., 16 June; The Cove, L. Huron.

Wild Yellow Lily. Lilium Canadense, L.
Ann Arbor; Ft. Gratiot; Sturgeon Pt., L. Huron, 30 June.

Turk's cap Lily. Lilium superbum, L.
Ann Arbor, (Miss Clark).

Yellow Adder's tongue. Erythronium Americanum, Smith.
Ann Arbor.

MELANTHACEÆ.

Large-flowered Bellwort. Uvularia grandiflora, Smith.
Ann Arbor.

Sessile-leaved Bellwort. Uvularia sessifolia, L.
S. E. (Wright).

Twisted-stalk. Streptopus amplexifolius DC.
Ft. Gratiot; St. Joseph's I.

Twisted-stalk. Streptopus roseus, Michx.
Drummond's I.; Sugar I.

Zygadene. Zygadenus glaucus, Nutt. (*Melanthium glaucum—W.)*
S. W. (Wright).

False asphodel. Tofieldia glutinosa, Willd.
False Presqu' Isle, L. Huron, 11 July; Drummond's I.

JUNCACEÆ.

Wood-rush. Luzula pilosa, Willd.
(Wright).

Wood-rush. Luzula compestris, DC.
Ann Arbor.

Common, or Soft Rush. Juncus effusus, L.
S. E. (Wright).

Rush. Juncus filiformis, L.
Saginaw Bay, 15 June.

Rush. Juncus Balticus, Willd.
Drummond's I., 25 July; Pine Lake, Emmet Co.; Leelanaw Co.; S. W. (Wright). Sandy shores, common.

Rush. Juncus setaceus, Rostk.
Sulphur I., north of Drummond's; S. Mich. (Wright).

Rush. Juncus paradoxus, E. Meyer. (*J. polycephalus—W.*)
S. Michigan, (Wright).

Rush. Juncus acuminatus, Michx.
S. Mich. (Wright).

Rush. Juncus articulatus, L.
Drummond's I., 22 July; Grand Traverse Bay (E. arm), abundant.

Rush. Juncus nodosus, L.
Drummond's I., 25 July; Grand Traverse Bay (E. arm), abundant.

Rush. Juncus marginatus, Rostk.
S. Mich., (Univ. Herb)

Rush. Juncus tenuis, Willd.
Sturgeon Pt., L. Huron; S. Mich. (Univ. Herb).

Rush. Juncus bufonius, L.
S. E. (Wright).

PONTEDERIACEÆ.

Pickerel-weed. Pontederia cordata, L.
Ann Arbor.

Water Star grass. Schollera graminea, Willd.
S. W. (Wright); Ann Arbor, (Miss Clark).

Day-flower. Commelyna Virginica, L. (*C. angustifolia—W.*)
S. W. (Wright).

Common Spider wort. Tradescantia Virginica, L.
Ann Arbor.

XYRIDACEÆ.

Yellow-eyed Grass. Xyris bulbosa, Kunth.
S. W. (Wright); Ann Arbor, (Miss Clark).

ERIOCAULONACEÆ.

Pipewort. Eriocaulon septangulare, Withering. (*E. pellucidum—W.*)
S. W. (Wright).

CYPERACEÆ.

Galingale. Cyperus diandrus, Torr.
Ann Arbor.
var. castaneus.
S. E. (Univ. Herb).

Cyperus flavescens, L.
S. Mich. (Wright).

Cyperus strigosus, L.
S. W. (Wright).

"Cyperus phymatodes, Muhl.?"
S. W. (Wright).

Cyperus filiculmis, Vahl. (*C. mariscoides—W.*)
S. Mich. (Wright).

Dulichium. Dulichium spathaceum, Pers.
S. Mich. (Wright).

Hemicarpa. Hemicarpa subsquarrosa, Nees. (*Scirpus subsquarrosa—W.*)
S. W. (Wright).

Spike-rush. Eleocharis equisetoides, Torr. (*Scirpus equisetoides —W.*)
S. E. (Wright).

Spike-rush. Eleocharis quadrangulata, R. Br.
S. Mich. (Gray).

Spike-rush. Eleocharis obtusa, Shultes. (*Scirpus capitatus—W.*)
S. E. (Wright).

Spike-rush. Eleocharis palustris, R. Br. (*Scirpus palustris—W.*)
Sturgeon Pt., L. Huron; Pt. au Chene, L. Mich.; S. Mich. (Wright).

Spike-rush. Eleocharis rostellata, Torr.
Drummond's I., 22 July.

Spike-rush. Eleocharis intermedia, Schultes.
Grand Traverse Bay.

Spike-rush. Eleocharis tenuis, Schultes. (*Scirpus tenuis—W.*)
S. E. (Wright).

Spike-rush. Eleocharis compressa, Sullivant.
Branch L., Emmet Co.

Spike-rush. Eleocharis acicularis, R. Br. (*Scirpus acicularis—W.*)
S. W. (Wright).

Bulrush. Scirpus subterminalis, Torr.
S. Michigan, (Wright).

Bulrush. Scirpus pungens, Vahl. (*S. Americanus—W.*)
Pt. au Chene, L. Mich., 19 Aug.; Grand Traverse Bay; S. Mich. (Wright).

Bulrush. Scirpus Torreyi, Olney.
Borders of ponds, (Gray).

Bulrush. Scirpus lacustris, L. (*S. lacustris and acutus—W.*)
Saginaw B., common; Pine Lake, Emmet Co., abundant; S. E. (Wright).
This species is extensively used by the Indians to make mats. It is cut late in summer just as the fruit is ripening. In Pine Lake it grows very large, the culm sometimes being 12 ft. or more in length.

Bulrush. Scirpus debilis, Pursh.
Low banks of streams, (Gray).

Sea Club-rush. Scirpus maritimus, L. (*S. macrostachyos—W.)*
S. Mich., (Wright).

River Club-rush. Scirpus fluviatilis, Gray.
S. Mich. (Univ. Herb).

Bulrush. Scirpus sylvaticus, L. var. atrovirens.
S. Mich. (Univ. Herb).

Bulrush. Scirpus polyphyllus, Vahl. (*S. brunneus—W.)*
S. Mich. (Wright).

Bulrush. Scirpus lineatus, Michx.
S. Mich. (Wright).

Wool-grass. Scirpus Eriophorum, Michx. (*S. Erisphoruo—W.?)*
S. W. (Univ. Herb).

Sheathed Cotton-grass. Eriophorum vaginatum, L.
S. Mich., (Univ. Herb).

Virginian Cotton grass. Eriophorum Virginicum, L.
S. Mich. (Univ. Herb).

Many stemmed Cotton grass. Eriophorum polystachyon, L.
S. Mich. (Wright).
Var. latifolium.
S. Mich. (Univ. Herb).

Graceful Cotton grass. Eriophorum gracile, Koch. (*E. angustifolium—W.*)
S. Mich. (Wright).

Fimbristylis spadicea, Vahl. (*Scirpus spadiceus—W.*)
S. W. (Wright).

Fimbristylis autumnalis, Roem. & Shult. (*Scirpus autumnalis—W.*)
S. Michigan, (Wright).

Fimbristylis capillaris, Gray. (*Scirpus capillaris—W.*)
S. Michigan, (Wright).

Umbrella-grass. Fuirena squarrosa, Michx.
S. Mich. (Wright).

Beak-rush. Rhynchospora alba, Vahl.
S. W. (Wright); N. E. (Univ. Herb).

Beak-rush. Rhynchospora capillacea, Torr.
Bogs and river banks, (Gray).

Beak-rush. Rhynchospora glomerata, Vahl.
S. Mich., (Wright).

Twig-rush. Cladium mariscoides, Torr. (*Schœnus mariscoides—W.*)
S. Mich. (Wright).

Nut-rush. Scleria triglomerata, Michx.
S. Mich., (Wright).

Nut-rush. Scleria verticillata, Muhl.
Swamps, (Cooley).

Carex gynocrates, Wormskiold.
N. E. and N. W., (Univ. Herb).

Carex scirpoidea, Michx.
N. E. (Univ. Herb).

Carex polytrichoides, Muhl.
S. Mich. (Wright); N. W. (Univ. Herb).

Carex bromoides, Schk.
Antrim Co.; S. E. (Wright).

Carex Sartwellii, Dew.
S. Mich. (Univ. Herb).

Carex teretiuscula, Good.
S. Mich. (Univ. Herb).

Carex decomposita, Muhl. (*C. paniculata—W.?*)
S. Mich. (Wright).

Carex vulpinoidea, Michx. (*C. setacea—W.*)
Sturgeon Pt., L. Huron; S. Mich. (Wright).

Carex stipata, Muhl.
S. Mich. (Wright).

Carex cephalophora, Muhl.
S. E. (Wright).

Carex rosea, Schk.
Ann Arbor.

Carex tenella, Schk. (*C. dispermia—W.*)
S. E. (Wright).

Carex trisperma, Dew.
(Wright).

Carex canescens, L. (*C. curta—W.*)
S. Mich. (Wright).

Carex Deweyana, Schw.
S. Mich. (Wright).

Carex stellulata, Good.
Sturgeon Pt., L. Huron; S. E. (Wright).
var, sterilis.
S. Mich. (Univ. Herb).

Carex scoparia, Schk.
S. E. (Univ. Herb).

Carex lagopodioides, Schk.
S. Mich. (Wright).
var. cristata, (*C. cristata—W.*)
S. Mich. (Wright).

Carex festucacea, Schk.
S. Mich. (Wright).
var. tenera, (*C. tenera—W.*)
S. Mich. (Wright).

Carex straminea, Schk.
S. Mich. (Wright).

Carex vulgaris, Fries. (*C. caespitosa—W.*)
S. Mich. (Wright).

Carex stricta, Lam. (*C. acuta—W.*)
S. Mich. (Wright).

Carex aquatilis, Wahl.
Near Sitting rabbit, 18 Aug.; S. E. (Wright).

Carex crinita, Lam.
Ann Arbor; Sturgeon Pt., L. Huron; S. Mich. (Wright)

Carex limosa, L.
S. Mich. (Wright).

Carex Buxbaumii, Wahl.
S. E. (Univ. Herb).

Carex aurea, Nutt.
S. E. (Wright).

Carex tetanica, Schk.
S. Mich. (Univ. Herb).

Carex Crawei, Dew.
N. Mich. (Bull).

Carex granularis, Muhl.
Drummond's I., 25 July; S. E. (Wright).

Carex conoidea, Schk.
S. E. (Wright).

Carex grisea, Wahl. var. mutica.
Drummond's I., 25 July.

Carex Davisii, Schw. & Torr.
Sitting rabbit.

Carex formosa, Dew.
S. Mich. (Wright).

Carex gracillima, Schk.
S. E. (Wright).

Carex virescens, Muhl.
S. Mich. (Wright).

Carex plantaginea, Lam. (*C. anceps—W.?*)
S. Mich. (Wright).

Carex laxiflora, Lam.
S. Mich., (Wright).

Carex eburnea, Booth. (*C. alba*, var. *setifolia*—*W.*)
Drummond's I., 28 July; L. Mich., Emmet Co.; S. W. (Wright).

Carex pedunculata, Muhl. (*C. lupulina*—.)
S. Mich. (Wright).

Carex Novæ-Angliæ, Schw. (*C. collecta, nigro-marginata*—*W.*)
S. E. (Wright).
var. Emmonsii.
Grand Traverse Bay.

Carex Pennsylvanica, Lam. (*C. marginata*—*W.*)
Ann Arbor.

"Carex varia, Muhl. ?"
S. Mich. (Wright).

Carex pubescens, Muhl.
S. Mich. (Wright).

Carex miliacea, Muhl.
S. Mich. (Wright).

Carex scabrata, Schw.
S. Mich. (Wright).

Carex arctata, Boott. (*C. sylvatica*—*W.*)
S. Mich. (Wright).

Carex flava, L.
Emmet Co.; S. E. (Wright).

Carex Œderi, Ehrh.
Drummond's I., 25 July.

Carex filiformis, Gmelin.
S. Mich. (Wright).

Carex languinosa, Michx.
S. E. (Univ. Herb).

Carex lacustris, Willa.
S. Mich. (Wright).

Carex aristata.
Lake shores and river-banks, (Univ. Herb).

Carex trichocarpa, Muhl.
S. Mich. (Wright).

Carex comosa, Boott.
S. Mich. (Univ. Herb).

Carex pseudo-cyperus, L.
S. Mich. (Wright).

Carex hystricina, Willd.
S. E. (Univ. Herb).

Carex tentaculata, Muhl.
Antrim Co.; S. Mich. (Wright).

Carex intumescens, Rudge.
N. W. (Univ. Herb.)

Carex folliculata, L. (*C. folliculata* and *xanthophysa—W.*)
S. Mich. (Wright).

Carex lupulina, Muhl.
Ann Arbor.

Carex squarrosa, L.
S. Mich. (Wright).

Carex retrosa, Schw.
S. Mich. (Wright).

Carex ampullacea, Good.
Bear Creek, Emmet Co.; S. Mich. (Wright).
var. utriculata.
S. E. (Univ. Herb).

Carex cylindrica, Schw.
S. Mich. (Univ. Herb).

Carex bullata, Schk.
S. Mich. (Wright).

Carex digosperma, Michx.
Oakland Co. (Prof. Williams.)

GRAMINEÆ.

Leersia oryzoides, Swartz.
S. Mich. (Wright).

White-grass. Leersia Virginica, Willd.
S. Mich. (Wright).

Indian Rice, Water Oats. Zizania aquatica, L.
(Wright).

Floating Foxtail. Alopecurus geniculatus, L.
S. Mich. (Wright).

Timothy, Herd's-grass. Phleum pratense, L.
Meadows, common.

Sporobolus cryptandrus, Gray.
S. Mich. (Univ. Herb).

Sporobolus serotinus, Gray.
Sandy wet places, (Gray).

Thin-Grass. Agrostis perennans, Tuckerm. (*Trichodium scabrum—W.*)
S. Mich. (Wright).

Hair-Grass. Agrostis scabra, Willd, (*Trichodium laxiflorum—W.*)
S. Mich. (Wright).

Brown Bent-grass. Agrostis canina, L.
Ann Arbor.

Agrostis vulgaris, With.
Ann Arbor.

White Bent-grass. Agrostis alba, L.
S. Mich. (Wright).

Cinna arundinacea, L.
S. W. Mich. (Wright).

Drop-seed Grass. Muhlenbergia sobolifera, Gray.
Open rocky woods, S. Mich. (Gray).

Drop-seed Grass. Muhlenbergia glomerata, Trin. (*Polypogon racemosus*—*W.*)
S. W. (Wright).

Drop-seed Grass. Muhlenbergia Mexicana, Trin. (*Agrostis lateriflora* —*W.*)
S. Mich. (Wright).

Drop-seed Grass. Muhlenbergia Willdenovii, Trin. (*Agrostis tenuiflora*—*W.*)
S. Michigan, (Wright).

Nimble Will. Muhlenbergia diffusa, Schreber.
S. Mich. (Wright).

Brachyelytrum aristatum, Beauv.
S. Mich. (Wright).

Blue Joint-Grass. Calamagrostis Canadensis, Beauv. (*Arundo Canadensis*—*W.*)
Pt. au Chene, L. Mich.; S. E. (Wright).

Reed Bent-grass. Calamagrostis coarctata, Torr. (*Arundo coarctata*—*W.*)
S. Mich., (Wright).

Calamagrostis longifolia, Hook.
Pt. au Chene, L. Mich., 19 Aug.; Antrim Co., common; S. W. (Univ. Herb).

Sea Sand Reed. Calamagrostis arenaria, Roth.
Pt. au Chene, L. Mich., 19 Aug.

Oryzopsis melanocarpa, Muhl. (*Piptatherum nigrum*—*W.*)
S. Mich. (Wright).

Mountain Rice. Oryzopsis asperifolia, Michx.
S. Mich. (Wright).

Oryzopsis Canadensis, Torr. (*Milium pungens*—*W.*)
S. E. (Wright).

Black Oat Grass. Stipa avenacea, L.
S. W. (Wright).

Porcupine Grass. Stipa spartea, Trin. (*S. juncea—W.*)
S. Mich., (Wright).

Aristida stricta, Michx.
S. Mich., (Wright). [Doubtful.]

Aristida purpurascens.
S. Mich., (Univ. Herb)

Fresh-Water Cord-Grass. Spartina cynosuroides, Willd.
S. Mich., (Wright).

Muskit-grass. Bouteloua curtipendula, Gray. (*Atheropogon apludioides—W.*)
S. Mich. (Wright).

Wire-grass. Elusine Indica, Gaertn.
S. Mich. (Wright).

Tall Red-Top. Tricuspis seslerioides, Torr.
S. W. (Wright).

Dupontia. Dupontia Cooleyi, Gray.
Washington, Macomb Co., (Gray).

Diarrhena. Diarrhena Americana, Beauv.
S. Mich. (Wright).

Kœleria. Kœleria cristata, Pers.
S. E. (Wright).

Eatonia obtusata, Gray. (*Kœleria truncata—W.*)
S. E. (Wright).

Eatonia Pennsylvanica, Gray. (*Kœleria Pennsylvanica—W.*)
S. Mich. (Wright).

Rattlesnake-Grass. Glyceria Canadensis.
S. W. (Univ. Herb.)

Glyceria elongata, Trin.
Wet woods, (Gray.)

Glyceria nervata, Trin. (*Poa nervata—W.*)
S. Mich. (Wright).

Reed Meadow-Grass. Glyceria aquatica, Smith. (*Poa aquatica,* var. *Americana—W.*)
S. Mich., (Wright).

Glyceria fluitans, R. Br.
S. Mich. (Wright).

Low Spear-Grass. Poa annua, L.
Ann Arbor; S. E. (Wright).

Poa debilis, Torr.
S. Michigan, (Univ. Herb).

Poa sylvestris, Gray.
S. Mich. (Univ. Herb).

False Red-Top, Fowl Meadow Grass. Poa serotina Ehrh.
Little Traverse Bay, 24 Aug.; S. Mich. (Wright).

Poa nemoralis, L.
S. Mich., (Wright).

Rough Meadow Grass. Poa trivialis, L.
S. Mich. (Wright).

Green, or Common Meadow Grass. Poa pratensis, L.
S. E. (Wright).

Blue-Grass, Wire-Grass. Poa compressa, L.
Ann Arbor.

Eragrostis reptans, Nees. (*Poa replans*—*W.*)
S. Mich. (Wright).

Eragrostis poæoides, (*Poa eragrostis*—*W.*)
Ann Arbor.

Eragrostis capillaris, Nees. (*Poa capillaris* and *hirsuta*—*W*)
S. Mich. (Wright).

Eragrostis pectinacea, Gray. *Poa hirsuta*—*W.*)
S. Mich. (Wright).
var. spectabilis.
S. Mich., (Univ. Herb).

Fescue-Grass. Festuca tenella, Willd.
S. Mich. (Wright).

Festuca ovina, Gray.
var. duriuscula. (*F. duriuscula*—*W.*)
S. Mich. (Wright).

Festuca nutans, Willd.
S. E. (Wright).

Cheat, Chess. Bromus secalinus, L.
Ann Arbor; Fields, Grand Traverse Co.

Wild Chess. Bromus Kalmii, Gray. (*B. ciliatus*—*W.*)
S. E. (Wright).

Bromus ciliatus, L.
Charlevoix, Emmet Co.; S. Mich. (Wright).
var. purgans, (*B. Purgans*—*W.*)

Reed. Phragmites communis, Trin.
S. Mich. (Wright).

Bearded Darnel. Lolium tremulentum, L.
S. Michigan, (Wright).

Couch-Grass, Quitch-Grass, Quick-Grass. Triticum repens, L.
S. Mich. (Univ. Herb).

Awned Wheat Grass. Triticum caninum, L. (*Agropyron caninum*—*W.*)
S. Mich. (Univ. Herb).

Triticum dasystachyum, Gray.
N. W. (Univ. Herb).

Elymus Virginicus, L.
S. E. (Wright).

Elymus Canadensis, L.
Drummond's I., 24 July; Antrim Co., common; S. E. (Wright).
var. glaucifolius.
S. Mich. (Univ. Herb).

Elymus striatus, Willd.
var. villosus. (*E. villosus—W.*)
S. Mich. (Wright).

Elymus mollis, Trin.
Shore of L. Huron, (Gray).

Bottle-brush Grass. Gymnostichum Hystrix, Schreb. (*Elymus Hystrix —W.*)
Ann Arbor; S. Mich. (Wright).

Hair-Grass. Aira cæspitosa, L.
S. Mich. (Wright).

Wild Oats. Danthonia spicata, Beauv.
S. Mich. (Wright).

Trisetum. Trisetum subspicatum, Beauv., var. molle, Gray.
N. E. (Univ. Herb).

Oat. Avena striata, Michx. (*Trisetum purpurascens—W.*)
S. E. (Wright).

Vanilla, or Seneca-Grass. Hierochloa borealis, Roem. & Schultes.
S. E. (Univ. Herb).

Reed Canary-Grass. Phalaris arundinacea, L.
S. Mich. (Univ. Herb.)

Millet-grass. Milium effusum, L.
S. E. (Wright).

Panic-Grass. Panicum filiforme, L. (*Digitaria filiformis—W.*)
S. W. (Wright).

Panicum glabrum, Gaudin.
Ann Arbor.

Crab-Grass, Finger-Grass. Panicum sanguinale, L. (*Digitaria sanguinale—W.*)
Ann Arbor.

Panicum capillare, L.
Ann Arbor; Mission Point, Grand Traverse Co.

Panicum virgatum, L.
S. Mich. (Wright).

Panicum latifolium, L.
S. Mich. (Univ. Herb).

Panicum dichotomum, L. (*P. nitidum* and *pubescens*—*W.*)
Grand Traverse Co.; S. Mich. (Wright).

Barnyard-Grass. Panicum Crus-galli, L.
Ann Arbor.

Panicum nervosum, Muhl.
S. Mich. (Wright). [A synonym?]

Foxtail. Setaria glauca, Beauv.
Ann Arbor.

Green Foxtail, Bottle-Grass. Setaria viridis, Beauv.
Ann Arbor; Emmet Co., growing in fields with the following.

Setaria Italica, Kunth.
Emmet Co., cultivated by the Indians as *Millet*.

Bur-Grass. Cenchrus tribuloides, L. (*C. echinatus*, var. *tribuloides*—*W.*)
S. W. (Wright).

Beard-Grass. Andropogon furcatus, Muhl.
S. Mich. (Wright).

Andropogon scoparius, Michx.
Antrim Co., common; S. E. and S. W. (Wright).

Andropogon Virginicus, L.
S. Mich. (Wright).

Broom-Corn. Sorghum nutans, Gray. (*Andropogon nutans*—*W.*)
S. Mich. (Univ. Herb.)

EQUISETACEÆ.

Horsetail, Scouring Rush. Equisetum arvense, L.
Ann Arbor; Bruce Mine, Ca.; Drummond's I., abundant, in sandy soil.

Meadow Horsetail. Equisetum pratense, Ehrh.
Ann Arbor; Pine Lake, Emmet Co.

Wood Horsetail. Equisetum sylvaticum, L.
Drummond's I.

Swamp Horsetail. Equisetum limosum, L.
Ann Arbor; S. E. (Wright).

Shave-Grass. Equisetum hyemale, L.
Ann Arbor; Drummond's I., very abundant in sandy soil; Branch Lake, Antrim Co., very abundant, growing in the marshy margin of the river near its mouth.

Scouring Rush. Equisetum variegatum, Schleicher.
Drummond's I.; S. E. (Univ. Herb).

Scouring Rush. Equisetum scirpoides, Michx.
Shore of Lake Michigan, Emmet Co.

FILICES.

Polypody. Polypodium vulgare, L.
Drummond's I.

Ostrich-Fern. Struthiopteris Germanica, Willd.
Ann Arbor.

Rock Brake. Allosorus gracilis, Presl.
Louse Island.

Rock Brake. Allosorus atropurpureus, Gray.
N. E. (Univ. Herb).

Common Brake. Pteris aquilina, L.
Ann Arbor; Ft. Gratiot; Drummond's I.; Emmet Co.; Traverse City; Ottawa Co.; S. Mich. (Wright). Common.

Maiden-hair. Adiantum pedatum, L.
Ann Arbor; Ft. Gratiot; Emmet Co., rich woods, common; S. Mich. (Wright).

Woodwardia. Woodwardia Virginica, Willd.
S. Mich. (Univ. Herb.)

Spleenwort. Asplenium Ruta-muraria, L.
N. E. (Univ. Herb).

Spleenwort. Asplenium Trichomanes, L.
N. E. (Univ. Herb).

Spleenwort. Asplenium angustifolium, Michx.
S. W. (Wright).

Silvery Spleenwort. Asplenium thelypteroides, Michx.
Ann Arbor; Ft. Gratiot; S. Mich. (Wright).

Spleenwort. Asplenium Filix-fœmina, R. Br.
Ann Arbor; Bear Creek, Emmet Co.; S. Mich. (Univ. Herb).

Dicksonia punctilobula, Hook.
Bear Creek, Emmet Co.

Woodsia. Woodsia Ilvensis, R. Br.
N. E. (Univ. Herb).

Bladder-Fern. Cystopteris bulbifera, Bernh. (*Aspidium bulbiferum*—*W.*)
Ann Arbor; S. E. (Wright).

Bladder-Fern. Cystopteris fragilis, Bernh.
Drummond's I.

Wood-Fern, Shield-Fern. Aspidium Thelypteris, Willd.
S. Mich. (Wright).

Wood-Fern, Shield-Fern. Aspidium noveboracense, Willd.
S. W. (Wright).

Wood-Fern, Shield-Fern. Aspidium spinulosum, Swartz. (*A. intermedium*—*W.*)
Ann Arbor; Emmet Co.; S. Mich. (Wright).
var. Bootii, Gray.
Ann Arbor.

Aspidium cristatum, Swartz.
Ann Arbor

Wood-Fern. Aspidium acrostichoides, Willd.
Ft. Gratiot; S. Mich. (Wright).

Wood-Fern. "Aspidium asplenoides, L."
S. Mich. (Wright).

Sensitive-Fern. Onoclea sensibilis, L.
Bear Creek, Emmet Co., 24 Aug.; S. W. (Wright); Ann Arbor.

Flowering Fern. Osmunda regalis, L.
Ann Arbor; Ft. Gratiot.
var. spectabilis.
Ann Arbor.

Osmunda Claytoniana, L. (*O. interrupta*—*W.*)
Ann Arbor; Ft. Gratiot.

Cinnamon Fern. Osmunda cinnamonea, L.
Ann Arbor; Ft. Gratiot.

Moonwort. Botrychium lunarioides, Swartz. (*B. fumarioides*—*W.*)
Ft. Gratiot; S. Mich. (Wright).

Botrychium Virginicum, Swartz.
Ft. Gratiot; Squaw Pt., Thunder Bay, 6 July; Drummond's I., 13 Aug.; Emmet Co., rich woods, rather common, S. Mich. (Wright).

LYCOPODIACEÆ.

Shining Club-moss. Lycopodium lucidulum, Michx.
Drummond's I., 24 July; Emmet Co.; S. E. (Wright).

Club-moss. Lycopodium inundatum, L. var. Bigelovii, Tuck.
Willow River, Huron Co., 20 June; Drummond's I.; Sugar I.

Club-moss. Lycopodium annotinum, L.
The Cove, L. Huron; Emmet Co., common.

Ground Pine. Lycopodium dendroideum, Michx.
Ft. Gratiot; Pt. aux Barques, Huron Co., 19 June; Sugar I., 31 July; N. W. (Univ. Herb).

Club-moss. Lycopodium clavatum, L.
Pt. aux Barques, Huron Co., 21 June; N. E. (Univ. Herb).

Club-moss. Lycopodium complanatum, L.
Traverse City, common in shade of pines; N. E. (Univ. Herb).

Selaginella apus, Spring.
Ann Arbor. This is not *S. selaginoides*.

CHAPTER X.

GENERAL REMARKS ON THE PRECEDING CATALOGUE.

Although the territory represented by the foregoing Catalogue does not extend into the Upper Peninsula, it nevertheless embraces a portion of the "Lake Superior Land District" as reported upon by the Botanist of Foster and Whitney's Survey. Within this portion of their territory, we have detected 95 species of plants not enumerated in W. D. Whitney's Catalogue.

The number of species embraced in this Catalogue is 274 more than in the Catalogue formerly published by Dr. Wright.

The total number of species enumerated (excluding varieties) is 1205. Of these, 85 species are of foreign origin. The introduced species embrace a large proportion of our common weeds The Black Mustard *(Sinapis nigra)*, Shepherd's Purse *(Capsella Bursa-pastoris)*, Mouse-Ear *(Cerastium vulgatum* and *C. viscosum)*, Purslane *(Portulaca oleracea)*, Mallows *(Malva rotundifolia)*, Corn Speedwell *(Veronica arvensis)*, Pigweeds *(Chenopodium hybridum* and *C. album)*, Amaranths *(Amarantus hybridus* and *A. retoflexus)*, Princes Feather *(Polygonum orientale)*, Sorrel *(Rumex acetosella)*, Crabgrass *(Panicum sanguinale)*, and the Foxtail grasses *(Setaria glauca* and *S. viridis)*, are common garden nuisances, and several of them spread themselves extensively through cultivated fields. The following more rarely encroach upon our gardens, but make themselves at home in cultivated and pasture fields: Buttercups *(Ranunculus acris)*, Horse Radish *(Nasturtium armoracia)*, Field Mustard *(Sinapis arvensis)*, Cockle *(Agrostemma Githago)*, Sandwort *(Arenaria serpyllifolia)*, Chickweed *(Stellaria media)*, Bladder Ketmia *(Hibiscus Trionum)*, White Melilot *(Melilotus alba)*, Common Daisy *(Lecanthemum vulgare)*, Groundsel *(Senecio vulgaris)*, Common and Canada Thistles *(Cirsium lanceola-*

tum and *C. arvense)*, Burdock *(Lappa major)*, Spiny Sow Thistle *(Souchus asper)*, Field Bindweed *(Convolvulue arvensis)*, Nighshade *(Solanum nigrum)*, Jamestown weed *(Datura stramonium)*, Wild Tobacco *(Nicotiana rustica)*, Lady's Thumb and Black Bindweed (*Polygonum Persicaria* and *P. convolvulus*), Hemp (*Cannabis sativa*), Brown Bent Grass (*Agrostis canina*), Floating Foxtail (*Alopecurus geniculatus*), Wire grass (*Eleusine Indica*), Eragrostis (*Eragrostis poœoides*), Chess (*Bromus secalinus*), and Barnyard grass (*Panicum crusgalli*) A few of our naturalized plants seem to have escaped from a state of cultivation, such as Red Clover (*Trifolium pratense*), Parsnep (*Pastinaca sativa*), Hyssop (*Hyssopus officinalis*), Peppermint (*Mentha piperita*), Horehound (*Marrubium vulgare*), Henbane (*Hyoscyamus niger*), Buckwheat (*Fagopyrum esculentum*) and Timothy Grass (*Phleum pratense*). Several species seem to be confined almost entirely to roadsides and waste places. Of such we may name Hedge Mustard (*Sisymbrium officinale*), Soapwort or Bouncing Bet (*Saponaria officinalis*), Cowherb (*Vaccaria vulgaris*), which is not common, Indian Mallow (*Abutilow Avicennœ*), equally rare, Spotted Hemlock (*Conium maculatum*), Wild Teasel (*Dipsacus sylvestris*), Elecampane (*Inula helenium*), Mayweed (*Maruta cotula*), Tansy (*Tanacetum vulgare*), Great Mullein (*Verbascum Thapsus*), an abundant pest in old fields, Toad Flax or Butter and Eggs (*Linaria vulgaris*), often a bold intruder into cultivated fields, Vervain (*Verbena hastata* and *V. urticifolia*), Catnep (*Nepeta Cataria*), Hemp Nettle (*Galeopsis tetrahit* and *G. Ladanum*), Motherwort (*Leonurus cardiaca*), Comfrey (*Symphytum officinale*), Gromwell (*Lithospermum arvense* and *L. officinale*), Stickseed (*Echinospermum Lappula*), Hound Tongue (*Cynoglossum officinale*), Apple of Peru (*Nicandra physaloides*), Jerusalum Oak and Mexican Tea (*Chenopodium botrys* and *C. ambrosioides*), Smartweed (*Polygonum hydropiper*), Dock (*Rumex crispus* and *R. obtusifolius*), and Stinging Nettle (*Urtica dioica*). But few trees and shrubs have been truly naturalized in the peninsula. Of such I have recognized the Sweet Brier (*Rosa rubiginosa*), very common on

Mackinac island, Bittersweet (*Solanum dulcamara*), the Brittle Willow (*Salix fragilis*), and the Lombardy Poplar (*Populus dilatata*).

A very considerable number of our wild plants are known to possess medicinal properties. Fourteen of the naturalized species fall into this category, viz: Toad Flax, Butter Cups, Black Mustard, Horse Radish, Spotted Hemlock, Elecampane, Bittersweet, (*Solanum dulcamara*), Jamestown Weed or Stramonium, Henbane, Great Mullein, Horehound, Peppermint, Wormseed, and Hemp. A more considerable number of our native plants hold an established place in the pharmacopœia, viz: Flowering Dogwood (*Cornus florida*), Spotted Cranesbill (*Geranium maculatum*), Butternut (*Juglans cinerea*), Mandrake (*Podophyllum peltatum*), Goldthread (*Coptis trifolia*), Black Snakeroot (*Cimcifuga racemosa*), Creeping Spearwort. (*Ranunculus flammula* var. *reptans*), Tulip tree (*Liriodendron tulipifera*), Bloodroot (*Sanguinaria Canadensis*), Seneca Snakeroot (*Polygala Senega*), Wood Sorrel (*Oxalis stricta*), Poison Ivy (*Rhus toxico dendron*), Indian Physic (*Gillenia trifoliata*), Wild Black Cherry (*Prunus serotina*), Ginseng (*Panax quinquefolium*), Dandelion (*Taraxacum dens-leonis*), Lobelia (*Lobelia inflata*), Wintergreen (*Gaultheria procumbens*), Bearberry (*Arctostaphylos uva-vrsi*), Prince's Pine (*Chimaphila umbellata*), Spice Bush (*Benzoin odoriferum*), Pleurisy Root (*Asclepias tuberosa*), Buckbean (*Menyanthes trifoliata*), Sassafras (*Sassafras officinale*), Hops (*Humulus lupulus*), Slippery Elm (*Ulmus fulva*), Juniper (*Juniperus communis*), Sweet Flag (*Acorus calamus*) Wild Turnip (*Arisœma triphyllum*), Columbo (*Frasera carolinensis*), which is different from the imported Columbo, Prickly Ash (*Zanthoxylum Americanum*), Agrimony (*Agrimonia eupatoria*), Fever Root (*Triosteum perfoliatum*), Black Alder (*Ilex verticillata*), Culver's Physic (*Veronica Virginica*), Pennyroyal (*Hedeoma pulegioides*), Dogbane (*Apocynum androsœmifolium*), Wild Ginger (*Asarum Canadense*), Pokeweed (*Phytolocca decandra*), Brake (*Pteris aquilina*), Wood Fern (*Aspidium Nove-*

boracense), Flowering Fern *(Osmunda regalis)*, Clubmoss *(Lycopodium clavatum)*. Several of the preceding are the American analogues of European species that enjoy, perhaps without reason, a greater reputation than the American ones. The American representatives of numerous other European species will undoubtedly be found to possess equal virtues with their foreign congeners; and not a few of these have already acquired considerable standing.

A number of our native plants, much larger than is generally supposed, are worthy of cultivation for ornament. Our peninsula affords some of the most magnificent shade trees known. The Sugar Maple *(Acer saccharinum)* has no superior, while the Silver Maple (*Acer dasycarpum*), Tulip tree (*Liriodendron tulipifera*), Basswood (*Tilia Americana*), Locust (*Robinia pseudo-acacia*), Kentucky Coffee Bean (*Gymnocladus Canadensis*), Honey Locust (*Gleditschia triacanthus*), Wild Black Cherry (*Prunus serotina*), Butternut (*Juglans cinerea*), Black Walnut (*Juglans nigra*), Balm of Gilead (*Populus balsamifera* var. *candicans*), and a number of others have long been extensively employed for shade and ornament. Besides these, our flora is rich in coniferous evergreens, of which the White Pine (*Pinus strobus*), Hemlock (*Abies Canadensis*), Balsam Fir (*Abies balsamea*), Black Spruce (*Abies nigra*), Arbor Vitæ (*Thuja occidentalis*), improperly called White Cedar, and Red Cedar (*Juniperus Virginiana*), are in greatest favor; while few trees offer a more graceful foliage than our Tamarack (*Larix Americana*). Of smaller sized ornamental trees may be mentioned the Hop Tree (*Ptelea trifoliata*), Striped Maple (*Acer Pennsylvanicum*) cultivated in Europe, Red Bud (*Cercis Canadensis*), Wild Crab Apple (*Pyrus coronaria*), Mountain Ash (*Pyrus Americana*), Flowering Dogwood (*Cornus florida*). Among shrubs ornamental in cultivation we have Stag's Horn Sumac (*Rhus typhina*), Burning Bush (*Euonymus atropurpureus*), Nine Bark (*Spiræa opulifolia*), Flowering Raspberry (*Rubus odoratus* and *R. Nutkanus*), Snow Berry (*Symphoricarpus racemosus*), Red

Berried Elder (*Sambucus pubens*) an attractive object at Mackinac and northward, Snowball (*Viburnum opulus*), Bear Berry (*Arctostaphylos uva-ursi*), Sheep Laurel (*Kalmia augustifolia*), which, with its beautiful and showy pink flowers, is very abundant at Thunder Bay, Trailing Red Cedar (*Juniperus Virginiana* var. *humilis*), Juniper (*Juniperus communis*), American Yew (*Taxus baccata* var. *Canadensis*). Of herbaceous plants attractive for the beauty of their flowers or the peculiarity of their foliage may be mentioned the Wild Columbine (*Aquilegia Canadensis*), more desirable than the foreign species, White Pond Lily (*Nymphœa odorata*) the various species of Violets, American Pitcher Plant (*Sarracenia purpurea*), Dodder (*Cuscuta Gronovii*), Sundew (*Drosera rotundifolia*), Fringed Polygala (*Polygala paucifolia*), Wild Lupine (*Lupinus perennis*), Goat's Rue (*Tephrosa Virginiana*), Silver Weed (*Potentilla anserina*), Great Willow Herb (*Epilobium augustifolium*), Evening Primrose (*Œnothera biennis*), Wild Valerian (*Valeriana sylvatica*), Blazing Star (*Liatris spicata*), Silky and Azure Asters (*Aster sericeus* and *A. azureus*), Compass Plant and Prairie Dock (*Silphium laciniatum* and *S. terebinthinaceum*), Cardinal Flower (*Lobelia cardinalis*), Syphilitic Lobelia (*Lobelia syphilitica*), Painted Cup *Castilleia coccinea*), Hairy Puccoon (*Lithospermum hirtum*), Moss Pink (*Phlox subulata*), Fringed and White Gentians (*Gentiana crinita* and *G. Alba*), Pleurisy Root (*Asclepias tuberosa*), Flowering Spurge (*Euphorbia corollata*) Showy Orchis (*Orchis spectabilis*), Large Round-leaved Orchis (*Platanthera orbiculata*), Yellow Fringed Orchis (*P. ciliaris*), Large Purple Fringed Orchis (*P. fimbriata*), Grass Pink (*Calopogon pulchellus*), Showy Lady's Slipper (*Cypripedium spectabile*), Turk's Cap Lily (*Lilium superbum*), Spiderwort (*Tradescantia Virginica*) and Maiden's Hair Fern (*Adiantum pedatum*). Among climbing and trailing plants may be mentioned, besides our native grapes and the trailing Bearberry, Red Cedar and Yew, our far famed American Ivy (*Ampelopsis quinquefolia*), our Virgin's Bower (*Clematis Virginiana*), the Climbing Bitter

Sweet (*Celastrus scandens*), and a delicate herbaceous vine, Climbing Fumitory (*Adlumia cirrhosa*) seen only on Middle Island of Lake Huron.

The Floras of the various sections of the peninsula are not yet sufficiently made known to justify any extended discussion of the geographical distribution of the species. Such facts as have been collected, however, foreshadow the nature of some general conclusions to which even now a brief reference may be made.

A large proportion of all our species are generally distributed, but the northern half of the peninsula receives a very considerable number of characteristic northern types. There is no definite line separating the boreal types from the austral, but in traveling northward we find a continual accession of forms more and more exclusively northern, until in the extreme northern limit of the district under consideration we find ourselves for the first time within the range of such species as *Primula farinosa, Mimulus Jamesii, Veronica alpina, Triglochin maritimum, var. elatum, Calypso borealis, Tofieldia glutinosa,* &c. A few species in that part of the district are almost or quite restricted to the White Mountains in their eastward distribution, while most of the others which characterize the northern district occur also in New York and Pennsylvania, and extend southward along the Alleghanies. It is worthy of particular remark that many of the species of Pennsylvania and New York are found in Michigan in a latitude considerably higher; while, in accordance with this fact, several of the species whose northern limit is in Ohio are found, further west, to have extended up into Michigan. The following are examples of species which, on a more easterly meridian, are not known to range as far north as our State: *Silene Pennsylvanica* (Wright), *Lespedeza repens* (Wright), *Cercis Canadensis, Agrimonia parviflora* (Miss Clark), *Liatris squarrosa, Rudbeckia speciosa, R. fulgida* (Miss Clark), *Vaccinium vacillans, Scutellaria integrifolia* (Wright), *Gentiana ochroleuca* (Miss Clark). A few more strictly Atlantic coast species, also, reappear in our State, mostly on a higher parallel than in their eastern habitat. Such are *Desmodium lœvigatum* (Wright), *D. strictum* (Wright), *Coreopsis trichosperma* (Wright), *Utricularia purpurea* (Wright), *Acnida cannabina* (Wright), *Bartonia tenella* (Wright), *Smilax tamnifolia*—though the appended authorities in these lists show that I have not generally verified the identifications. It would seem then that the isofloral lines, like the isothermal ones, are, in their westward prolongation, deflected somewhat toward the north, though the deflection is considerably more in the former than the latter.

The following is a list of the species which have not been observed south of the mouth of the Saginaw river. It cannot by any means be asserted, however, that none of these occur in the more southern counties, though very few, if any, will be discovered as far south as Ann Arbor:

List of Native Plants not observed south of the mouth of Saginaw river.

Anemone multifida,
Corydalis aurea,
" glauca,
Sisymbrium arabidoides,
Turritis glabra,
" stricta,
Barbarea vulgaris,
Sisymbrium canescens,
Cakile Americana,
Viola rotundifolia,
Hudsonia tomentosa,
Drosera rotundifolia,
Geranium Robertianum,
Acer Pennsylvanicum,
Acer spicatum,
Rubus Nutkanus,
Pyrus Americana,
Amelanchier Canadensis,
vars. botryapium & alnifolia,
Epilobium palustre, var. liniare,
Ribes lacustre,
" prostratum,
Lonicera parviflora,
" hirsuta,
" ciliata,
Nardosmia palmata,
Aster simplex,
Solidago puberula,
" stricta,
" Houghtonii,
Coreopsis lanceolata,
Tanacetum Huronense,
Artemesia Canadensis,
" Ludoviciana,

var. gnaphalodes,
Antennaria Margaritacea,
Cirsium Pitcheri,
" undulatum,
Hieracium Canadense,
Chiogenes hispidula,
Kalmia angustifolia,
" glauca,
Ledum latifolium,
Pterospora Andromeda,
Primula farinosa,
Mimulus Jamesii,
Veronica Alpina,
Gerardia aspera,
Halenia deflexa,
var. linearis,
Blitum capitatum,
Polygonum articulatum,
" cilinode,
Rumex altissimus,
" salicifolius,
Corylus rostrata,
Betula papyracea,
Betula lenta,
Alnus incana,
Populus balsamifera,
Pinus Banksiana,
" resinosa,
Abies Fraseri,
" alba,
Juniperus Virginiana,
var. humilis,
Taxus baccata, var. Canadensis,
Potamogston pectinatus,
" prælongus,
Triglochin maritimum,
var. elatum,
Goodyera repens,
Calypso borealis,
Tipularia discolor,
Corallorhiza Macræi,
Iris lacustris,
Trillium erythrocarpum,
Smilacina trifolia,
Streptopus roseus,
Tofieldia glutinosa,
Luzula pilosa,
Juncus filiformis,
" Balticus,
" articulatus,
" nodosus,
Eleocharis rostellata,
" intermedia,
Carex gynocrates,
" scirpoidea,
" trisperma,
" Crawei,
" Œderi,
" grisea,
" aristata,
Zizania aquatica,
Calamagrostis arenaria,
Oryzopsis asperifolia,
Poa serotina,
Triticum dasystachyum,
Elymus mollis,
Aira caespitosa,
Trisetum subspicatum,
Equisetum sylvaticum,
" scirpoides,
Allosorus atropurpureus,
Asplenium Ruta-muraria,

Asplenium Trichomanes,
Woodsia Ilvensis,
Cystopteris fragilis,
Lycopodium inundatum,
var. Bigelovii,
Lycopodium annotinum,
" complanatum.

Future observations will undoubtedly greatly reduce the foregoing list, as well as the following:

List of native Plants seen only on the southwestern slope of the Peninsula.

Amorpha canescens,
Desmodium canescens,
Lespedeza violacea,
var. augustifolia,
Lespedeza hirta,
Ludwigia alternifolia,
Chrysoplenium Americanum,
Hydrocotyle umbellata,
Eryngium yuccæfolium,
Thaspium barbinode,
" trifoliatum,
Vernonia fasciculata,
Liatris spicata,
Solidago ulmifolia,
Silphium laciniatum,
" integrifolium,
Echinacea purpurea,
Helianthus occidentalis,
Hieracium Gronovii,
Lysimachia lanceolata,
var. hybrida,
Mimulus alatus,
Veronica anagallis,
Buchnera Americana,
Gerardia auriculata,
Scutellaria pilosa,
Cuscuta Gronovii,
Bartonia tenella,
Bœhmeria cylindrica,
Celtis occidentalis,
var. crassifolia,
Triglochin palustre,
Scheuchzeria palustris,
Sagittaria pusilla,
Gymnadenia tridentata,
Spiranthes gracilis,
Pogonia ophioglossoides,
" pendula,
Microstylis ophioglossoides,
Liparis liliifolia,
Zygadenus glaucus,
Commelyna Virginica,
Eriocaulon septangulare,
Cyperus strigosus,
" phymatodes,
Hemicarpha subsquarrosa,
Eleocharis acicularis,
Fimbristylis spadicea,
Agrostis scabra,
Muhlenbergia glomerata,
Stipa avenacea,
Tricuspis seslerioides,
Glyceria Canadensis,
Panicum filiforme,

Cenchrus tribuloides, Aspidium Noveboracense.
Asplenium augustifolium,

At Stone island and Drummond's island some pains were taken to make out pretty extended lists of the plants noticed. Stone island is the middle one of three small islands in Saginaw Bay, lying near the east shore. The following species were noted at these two localities.

1.—*Vegetation of Stone Island, Saginaw Bay.*

Pinus Strobus, Thuja occidentalis, Tilia Americana, Pteris aquilina, Geranium Robertianum, Actæa spicata, Trillium erectum, Smilacina bifolia, Ribes cynosbati, Galium circæzans, Cratægus coccinea ——? Rhus glabra, R. Toxicodendron, Zanthoxylum Americanum, Rhus typhina, Erigeron Philadelphicum, Aquilegia Canadensis, Sassafras officinale, Vitis cordifolia, Quercus tinctoria, Smilax ——? Geranium maculatum, Prunus ——, Achillea millefolium, Viola cucullata, Eupatorium perfoliatum, Anemone Pennsylvanica, Fragaria Virginiana, Rubus (small vine), Galium trifidum, Ranunculus abortivus, Erigeron Philadelphicum, Rubus villosus, Podophyllum peltatum, Sanicula Canadensis, Ribes floridum, Carpinus Americana, Hypoxys erecta, Cratægus tomentosa, var. mollis, Potentilla Canadensis, Acer saccharinum, Acer nigrum, Potentilla anserina, Castilleia coccinea, Apocynum androsæmifolium, Rosa blanda, Calystegia spithamæa, Nabalus ——, Iris versicolor, Polygala senega, Brunella vulgaris, Stellaria longifolia, Turritis stricta, Heracleum lanatnm, Thalictrum cornuti, Cornus stolonifera, Cornus paniculata, Linaria Canadensis, Cypripedium pubescens, Antennaria plantaginifolia.

2.—*Flora of Drummond's Island.*

Cirsium undulatum, Lonicera parviflora, Platanthera orbiculata, Abies alba, Actæa spicata, var. alba, Castilleia coccinea, Lycopodium clavatum, Platanthera dilatata, Hypericum prolificum, Brunella vulgaris, (a variety with white corolla,) Eupatorium perfoliatum, Calamintha glabella, var. Nutallii, Usnea barbata, Parnassia palustris, Lycopus Europæus, var. sinuatus, Arctostaphylos Uva-ursi, Primula farinosa, Solidago Houghtonii, Solidago stricta, Platanthera psycodes, Spiranthes latifolia, Eleocharis rostellata, Pteris aquilina, Campanula rotundifolia, very abundant, Juncus articulatus, Anemone Virginiana, Botrychium Virginicum, Alnus incana, abundant, Spiræa opulifolia, common, Rosa lucida, Thuja occidentalis, Larix Americana, Abies balsamea, Pinus resinosa, Acer saccharinum, Fagus ferruginea, Populus tremuloides, Clintonia borealis, Quercus rubra, Corydalis aurea, Cornus stolonifera, Fragaria Virginiana, Cornus circinata, Betula papyracea, Epilobium angustifolium, Geranium Carolinianum, Blitum capitatum, Polygonum cilinode, Pinus strobus, Acer spicatum, Acer Pennsylvanicum, Rubus triflorus, Taxus baccata, var. Canadensis, Aralia nudicaulis, Diervilla trifida, Cornus Canadensis, Chimaphila umbellata, Rhus toxicodendron, Rumex acetosella, Amelanchier Canadensis, Corydalis glauca, Rosa blanda, Salix candida, Salix lucida, Epilobium coloratum, Potentilla fruticosa, Salix pedicellaris, Smilacina racemosa, Lonicera hirsuta, Physalis viscosa, Ribes lacustris, Lycopodium inundatum, Lycopodium lucidulum, Melampyrum Americanum, Œnothera biennis, Achillea millefolium, Geum strictum, Lonicera parviflora, Ostrya Virginica, Tilia Americana, Erigeron Canadense, Symphoricarpus racemosus, Sambucus pubens, Chenopodium hybridum, Aster cordifolius, Potentilla Norvegica, Blephilia ciliata, Ulmus Americana, Sanicula Marilandica, Anemone multifida, Prunus Virginiana, Fraxinus Americana, Betula lenta, Prunus pumila, Cornus Canadensis, Linnæa borealis, Abies nigra, Juniperus communis, Juniperus Virginiana, var. humilis, Populus balsamifera, Gaylussacia resinosa, Spirea salicifolia, Comandra umbellata, Triglochin maritimum, var. elatum, Viola cucullata, Brunella vulgaris, Senecio aureus, var. balsamitæ, Polygala senega, Iris lacustris, Potentilla anserina, Ribes hirtellum, Eupatorium purpureum, Tofieldia glutinosa, Lilium Philadelphicum, Antennaria margaritacea, Zanthoxylum Americanum, Anemone Pennsylvanica, Ribes hirtellum, Vitis cordifolia, Trillium grandiflorum, Elymus Canadensis, Cornus circinata, Geranium Robertianum, Salix humilis, Lathyrus palustris, Salix sericea, Juncus nodosus, Salix discolor, Salix eriocephala, Juncus Balticus, Equisetum sylvaticum, Equisetum variegatum, Carex granularis, Lobelia Kalmii, Carex eburnea, Solidago Canadensis, Solidago lanceolata, Carex grisea, var. mutica, Carex Œderi, Erigeron Philadelphicum, Polygala paucifolia, Cypripedium pubescens, Fraxinus pubescens, Hieracium Canadense, Hypericum Canadense, Solidago puberula, Solidago Ohioënsis, Erigeron strigosum, Aster ericoides, Erigeron Canadense, Mulgedium leucophæum, Nepeta cataria, Rumex crispus, Aralia racemosa, Actæa spicata, Aster ——? Naumburgia thyrsiflora, Aster ptarmicoides, Abies Canadensis, Populus grandidentata, Lappa major, Aster cordifolius, Abies balsamea, Pyrola elliptica, Coreopsis lanceolata, Lathyrus maritimus, Lycopodium clavatum, Apocynum androsæmifolium, Aster sagittifolius, Equisetum hyemale, Equisetum arvense, Pteris aquilina, Sanguinaria Canadensis, Corylus rostrata, Cirsium muticum, Pastinaca sativa, Galeopsis Tetrahit, Stellaria longifolia, Mentha Canadensis, Eupatorium perfoliatum, Ribes floridum, Populus dilatatus, Lathyrus palustris, Gentiana detonsa, Solidago nemoralis, Cystopteris fragilis, Ranunculus abortivus, Artemisia Canadensis, Campanula aparinoides, Polypodium vulgare. Total observed, 189 species.

INDEX.

INDEX.

A.

B.

www.ingramcontent.com/pod-product-compliance
Lightning Source LLC
LaVergne TN
LVHW010206110826
845151LV00002B/623
9781425535100